INTRODUCTION TO PASCAL FOR SCIENTISTS

Introduction to Pascal for Scientists

BY JAMES W. COOPER
Vice-President for Software Development
Bruker Instruments, Inc.
Billerica, Mass 01821

A WILEY-INTERSCIENCE PUBLICATION
JOHN WILEY & SONS New York · Chichester · Brisbane · Toronto

Cooper, James William, 1943-
Introduction to PASCAL for scientists.
"A Wiley-Interscience publication."
Includes bibliographies and index.
1. PASCAL (Computer program language) I. Title.
QA76. 73. P2C68 001.64'24 80-28452
ISBN 0-471-08785-8

Printed in the United States of America

10 9 8 7 6 5 4 3 2 1

Preface

This text on Pascal was written to provide a simple guide for the scientist interested in the Pascal language. Early texts were aimed at experienced programmers and compiler writers, and the more recent teaching texts have not been directed at the scientist in particular.

In the present work we present fully working programs in all examples, rather than program fragments. Each program has been tested and runs as shown on one or more Pascal compilers. Further, we have commented these programs much more thoroughly than other authors since we feel that the comment features of Pascal are one of its strongest points and that no program can be easily understood without a large number of comments.

In this text we refer to three main compilers: DEC-10 Pascal, UCSD Pascal, and Aspect Pascal. DEC-10 Pascal was developed at Carnegie–Mellon University from the Hamburg compiler of Kisicki and Nagel. The UCSD Pascal compiler was based on the portable p-code compiler developed at the ETH in Zurich and was further developed under the leadership of Dr. Kenneth Bowles at the University of California–San Diego. The name "UCSD Pascal" is a trademark of the Regents of the University of California, and the compiler is available under license from Softech Microsystems (San Diego, California) for the PDP-11, Intel 8080, Motorola 6800, 6502, and Texas Instruments 9900.

Aspect Pascal is an interpreter for the Pascal p-code compiler developed to run under the ADAKOS operating system on the Aspect-2000 minicomputer made by Bruker Instruments. This text is based on a manual that I originally wrote for Bruker Instruments, Inc., describing Aspect Pascal, and their permission to use that material is gratefully acknowledged.

In this text I have intentionally used different terminology than Wirth in a few places to keep the presentation as clear to the novice as possible. In particular, I have used "calling parameter" for "actual-parameter" and "dummy parameter" for "formal-parameter." I am also fully aware that my use of the word "list" is nonrigorous from the computer-science point of view. I have omitted a few of the newer features of Pascal in the interest of clarity, such as the variant forms of NEW and DISPOSE and conformant arrays.

The "proposed standard" mentioned throughout the text is the draft proposal being considered by the ISO standards committee. This proposal is still under discussion and has not yet been approved, but it comes closest to a standard that can be cited. Special thanks go to A. Winsor Brown of Point-4 Data Corporation, a member of the ISO standards committee, for his thoughtful review of this manuscript.

I would like to thank all of the students in my Tufts Pascal course, as well as Professor Robert Stolow and Leo Joncas for reading the manuscript and commenting on it. I would especially like to acknowledge the meticulous review given the work by Dr. Rolf Johannesen of the National Bureau of Standards.

Finally, I would like to acknowledge the cooperation of the Tufts University Computer Center and the help of George Stalker and Doug Rayner in obtaining and installing the Carnegie–Mellon University (CMU) Pascal compiler.

James W. Cooper

Billerica, Massachusetts
March 1981

Contents

INTRODUCTION TO PASCAL FOR SCIENTISTS

CHAPTER 1

Introduction

The Pascal language was developed in the late 1960s by Niklaus Wirth at the ETH in Zurich; it was derived primarily from ALGOL-60. Since the language is named for the French mathematician Blaise Pascal, it is usually not represented in capital letters as if it were an acronym. Pascal is a language specifically designed to facilitate *structured* programming or top-down programming —a technique that has led to great efficiencies in program writing and maintainence. Pascal has thus been described as "Blaising" new trails in structured programming.

The Pascal language has recently achieved great popularity as more and more people discover how easy it is to write clear bug-free programs that are easily read and maintained. Some of the features of Pascal that make this possible include:

1 Several structured looping commands.

2 The CASE statement to allow many-branched decisions.

3 Simple input and output specifications.

4 Manipulation of sets, records, and linked lists.

5 Ease of handling and comparing strings and characters.

6 A comment structure that allows comments to occupy many lines and appear anywhere within a line or program.

7 Free format, which greatly improves readability.

In addition, Pascal has a large complement of features for arithmetic and logical manipulations, including some that make decision-making statements extremely simple to write. In short, Pascal has all the power of FORTRAN or PL/1, but much greater structure, and is, in fact, only slightly more complex to learn than BASIC, making it the ideal language for both beginners and experienced programmers.

The term *compiler* refers to a program that translates statements in a complex language such as Pascal, FORTRAN, BASIC, COBOL, or PL/1 into simple machine instructions which a computer can actually perform, one step at a time. Each different model of computer will probably have a different set of fundamental operations that it can perform, and thus each such computer will have a different compiler written for it for each high-level language that it has available.

For example, the Pascal statement

```
Y:=M*X + B;
```

might be translated by some compiler into the following fundamental operations:

```
FETCH value in memory location reserved for X
MULTIPLY by value in memory location reserved for M
ADD value in memory location reserved for B
STORE result in memory location reserved for Y
```

One of the features of Pascal that grows out of its structured approach to programming is the fact that a *one-pass* compiler can easily be written to translate Pascal programs into machine language. Since all constants, variables, and routines are declared in advance at the beginning of the program, not only can the programmer or user easily determine their meaning and use, but the compiler can as well. Thus Pascal compilers need only scan through the text once and are quite fast and efficient compared to those of most other languages.

A compiler that translates the Pascal (or other language) statements into machine language directly is the most common type and is the least "portable" since it will not operate on any other make of computer. A useful compromise has been developed by Wirth and colleagues and refined by the UCSD Pascal project, where the compiler translates the statements not into the code for any specific machine but into instructions for a hypothetical computer or *p-machine.* This *p-code* can then be executed by simply writing a program to execute each of the instructions of the hypothetical machine's instruction set. This is a *p-code interpreter,* and such interpreters provide relatively fast execution of Pascal programs for a number of minicomputers and microprocessors. Such p-code programs also occupy a minimum of space.

The standards for a Pascal compiler are being developed from the Pascal report by Jensen and Wirth (1) and from additional documents published by them. These standards (2) specify a very sparse language with a number of intentional omissions, but which contains the tools to allow these omitted features to be easily added. Thus while "standard Pascal" is quite simple, a number of additions have occurred since, which make the various compilers slightly different in features. We sometimes refer to these compilers as translating different "dialects" of Pascal.

When Pascal was first developed, time-sharing computers were not too common; thus Pascal was originally a batch-oriented language, where the program and its associated data were punched on cards and submitted for running. The output was then picked up later a the computer center. Since more and more computers support and encourage interactive terminal use rather than batch use, and since most minicomputers work exclusively in this mode, we will discuss the interactive terminal use of Pascal whenever possible. We will point out the virtues of this approach and the necessity for good conversational programming in such systems.

PROBLEMS

1 What does a compiler do?

2 What is a portable compiler?

3 What advantages are there to a one-pass compiler?

REFERENCES

1 K. Jensen and N. Wirth, "Pascal User Manual and Report," 2nd edition, Springer-Verlag, New York, 1976.

2 ISO Draft Proposal 7185 (ISO/TC 97/SC 5 N 565); *Pascal News,* 18, May 1980. Also avalable on request from X359 Secretary, c/o ×3 Secretariat, CBEMA: Suite 1200, 1828 L St. NW, Washington, D.C. 20036.

CHAPTER 2

A Simple Pascal Program

A LOOK AT THE PROGRAM

Let us take a look at a simple program for adding two numbers together and examine the fundamental features of Pascal in detail. The program below adds two numbers and prints the result on the terminal:

```
PROGRAM SUMMER;
CONST
     A=5.6;
     B=9.8;
VAR
     C:REAL;
BEGIN
     C:=A+B;
     WRITELN('THE SUM OF A AND B IS ',C);
END.
```

This program is not too hard to read for an experienced Pascal programmer, but may be baffling to the uninitiated. Thus as our first Pascal feature, we shall introduce the *comment.*

Comments may occur anywhere in a Pascal program where a space may appear and must be enclosed in braces or, on some systems, in a parenthesis-star pair:

```
(*THIS IS A COMMENT*)
{THIS IS A COMMENT IN BRACES}
```

Since braces do not appear on many commonly used terminals, we will use the parenthesis-star pair exclusively in this text.

The proposed standard allows comments to begin with one type of comment character and end with the other, but many Pascals allow only one of the two types, and UCSD Pascal requires that a comment begin and end with the same type of comment character to allow for comment nesting. Comments are totally ignored by the Pascal compiler itself, but are printed out in the listing for the use of the programmer and others who may want to see how the program works. Comments may be many lines long and may occur even in the middle of statements, but they must start and end eventually with the parenthesis-star pair.

While comments have no effect on the compilation or execution of the program, they are vital to any well-written program, since they provide information:

1 To the programmer, who after a few weeks may have forgotten how the program works in detail

2 To other users or programmers who may take over the use or maintenance of the program when the original programmer goes on to another project

3 To potential users at other locations

A well-commented program (in any language) is a valuable tool; an uncommented program is almost unreadable and is thus useless garbage. Consequently, many instructors in programming courses give zero credit for uncommented programs. It is essential to learn the habit of commenting programs as you type them in, since the activation energy to enter comments after the program is working is much too high to overcome.

Now let us look at this same program with some comments added:

```
PROGRAM SUMMER;
(*PROGRAM TO ADD  A AND B AND PLACE THE RESULT IN C *)
CONST              (*DECLARE THE CONSTANTS *)
     A=5.6;
     B=9.8;
VAR                (*DECLARE THE VARIABLES *)
     C:REAL;       (*C IS OF TYPE REAL*)

BEGIN              (*MAIN PROGRAM BEGINS HERE*)
     C:= A+B;      (*ADD  A AND B, PUT THE RESULT IN C*)
     WRITELN('THE SUM IS',C) (*PRINT OUT THE SUM*)
END.               (*PROGRAM ENDS HERE. STOP WHEN DONE*)
```

This simple program illustrates the major points of the structure of all Pascal programs:

1 All programs must start with a PROGRAM header which is at least followed by the name of the program.

2 Constants in the program may be declared in advance, by name. Such a group of constants starts with CONST. The actual value of the constant is given in the declaration, preceded by its name and an equal sign. (Numerical constants may also appear directly in the program code.)

3 Variables must likewise be declared in advance, and must be assigned a type. Such a group of variables must start with VAR, which must follow the CONST section. Variables of the same type must be listed separated by commas and terminated by a colon and a type name such as REAL.

4 The executable part of a program must start with BEGIN.

5 All programs must end with END followed by a period.

6 All statements must be separated with *semicolons* (;). The last statement has nothing it needs to be separated from and thus, strictly speaking, need not be followed by a semicolon. However, in most Pascal systems these

semicolons do no harm, and we will include them in our programs in this text for simplicity.

7 An assignment statement means *store* the result of a calculation into a memory location reserved for a variable. The variable on the left-hand side of the equation is separated from the expression on the right by a colon and an equal sign (:=). Thus the statement "add the values of A and B together and store the result in C" is written

```
C:=A+B;
```

8 Comments may appear anywhere in a program—on the same line, on separate lines, and even in the middle of expressions.

9 The output statement WRITELN (list) will print out both messages ('THE SUM IS') and values of variables (C).

COMPILER COMMENTS

Many Pascal compilers recognize a $ sign just inside the comment as a special signal to the compiler, called a *compiler comment.* The specifics of these comments vary with each system, and local manuals should be consulted.

For the most part, these comments are of the form:

```
(*$c+ *)      or      (*$c- *)
```

where c is a single character specifying a compiler function which is to be turned on (+) or off (−). For example, the comment

```
(*$L+ *)
```

in the UCSD Pascal system turns on the listing option, and the comment

```
(*$G+ *)
```

allows the use of the controversial GOTO statement.

SPACING AND LINES

We could actually write the same program in an even more confusing way:

This program, in addition to being uncommented, is strung together on a few lines, so that it becomes almost unreadable. It does, however, illustrate one important advantage of Pascal:

> New Lines (carriage returns) and spaces may be inserted in any place where a blank may be to improve program readability. They are ignored by the compiler.

We thus adopt the convention of starting new program parts on new lines and *indenting* subsections further. We then insert comments wherever needed to improve readability—either setting off a series of statements or following each statement on the same line. Comments can go on for several lines, as long as they eventually end with a right brace or a star-right parenthesis.

UPPERCASE AND LOWERCASE

For terminals that print in both uppercase and lowercase, Pascal will recognize both capital and lowercase characters as equivalent in statements. Any kind of characters may, of course, appear in comments. Thus

```
A:=B+C;      and      a:=b+c;
```

mean exactly the same thing, regardless of whether A, B, and C were originally declared using uppercase or lowercase characters. Most Pascal systems allow both uppercase and lowercase in data read in during program execution and in messages to be printed out. Since not all terminals and systems have lowercase available, we will show nearly all programs in this book in uppercase. This also provides some contrast with the surrounding text material.

INPUT AND OUTPUT—READ AND WRITE

In order to make our simple addition program more versatile, we would like to be able to read in the values to be added together, thus making A, B, and C variables. Entry of data can be accomplished with the READ statement:

```
READ(A, B);
```

Unless otherwise specified, the data will be read from the terminal in minicomputer or time-sharing systems, and from an input deck in batch systems.

In DEC-10 Pascal and UCSD Pascal all READs are done from the terminal and all WRITEs are done to the terminal unless another device is specified, as shown in Chapter 9. In most Pascals the device name INPUT is implied if no other device is specified in READ statements:

```
READ (INPUT,A,B);
```

or

```
READ (A,B);
```

and the device name OUTPUT implied in all WRITE statements:

```
WRITE(A,B);
```

or

```
WRITE(OUTPUT,A,B);
```

Data is read by the READ statement until a terminating character is encountered for that data type. For real numbers, terminating characters include spaces, tabs, or carriage returns. This terminating character then remains the next character to be read in a future READ statement.

By a *real* number we mean one that has a fractional part. Real numbers can be entered in any standard decimal format with or without leading spaces, signs, or zeros. Thus we can represent 235 in any of the following forms:

```
+235.0
235.0
235.000
2.35E2
2350.0E-1
```

where the number following the E means the power of 10 by which the value is to be multiplied (*not* the power of e!). Standard Pascal requires that a real number include a digit on both sides of the decimal point even if it must be a zero, but a few systems are actually more forgiving than this. The proposed standard allows the entry of exponential numbers without decimal points. Thus 1200 can be entered as 12E2 or as 12.0E2.

It is not necessary that each individual number be on a separate line: as many numbers as desired may be entered on the same line. However, for clarity you may require that a carriage return be typed by using the READLN statement, which ignores all characters until a new line is started with a return. In batch card-oriented systems, READLN skips to the next card or the next line of the input file.

Thus if we wanted to enter A and B on the same line and then require data on a new line, we *could* write:

```
READ(A,B);
READLN;
```

This can be abbreviated, however, by following the READLN statement with a list of the data to be read.

```
READLN(A,B);
```

is thus entirely equivalent to

```
READ(A,B);  READLN;
```

In a similar manner we can also print out the values of variables and strings of characters. The WRITE statement prints out all data inside the parentheses on a single line, and the WRITELN statement prints out all the data on a single line and then prints a return. The statement

```
WRITE(A,B);
```

prints out the values of A and B as real numbers if A and B were declared REAL and as integers if they were declared INTEGER. The statement

```
WRITELN(A,B);
```

prints out the same values, followed by a return, thus starting a new line for further output. A thorough discussion of how to format numerical output and other details of input and output is given in Chapter 7.

STATEMENTS AND DECLARATIONS

We distinguish between Pascal statements and declarations as follows:

A *statement* contains executable instructions—it tells Pascal to perform some operation. These are statements:

```
C:=A-B;
Y:=M*X + B;
READ(A,B);
```

Pascal *declarations* are primarily instructions to the compiler to help in translating the Pascal statements into machine code. They cannot, in themselves, be executed, but they tell the compiler how to translate statements that follow. Examples of declarations are:

```
PROGRAM ADDER;
CONST X=5.2;
VAR A,B,C: REAL;
```

Pascal *compound statements* (or *blocks*) are simply groups of statements that are always executed together. This allows grouping of statements that logically belong together. One of the great strengths of Pascal is that any statement can be replaced by a compound statement. Such compound statements are set off from the remainder of the program by the words BEGIN and END. A compound statement might be:

```
BEGIN
   Z:=A+B;
   Y:=M*X + B;
END;
```

Further, compound statements themselves may be made up of compound statements to almost any level of nesting, as we shall see in future chapters.

SEMICOLONS, CARRIAGE RETURNS, AND SPACES

Pascal ignores carriage returns and spaces anywhere in a program. Thus several short statements or declarations may appear on one line, and a long statement or declaration may occupy several lines. Since carriage returns are not significant, statements are separated by semicolons. This is not strictly the same as having semicolons terminate statements, since semicolons need not appear at the end of a block of statements where there are no statements to separate. However, using semicolons in such places has little effect, implying an "empty statement," but this usually does not affect program execution time at all. For example,

```
BEGIN
   Z:=A+B;          (*FIRST STATEMENT OF BLOCK*)
   Y:=M*X+B         (*LAST STATEMENT, NO ";" NEEDED*)
END;
```

but

```
BEGIN
   Z:=A+B;
   Y:=M*X+B;        (*EMPTY STATEMENT IMPLIED AFTER ";" *)
END;
```

In the second compound statement above, the second semicolon is not needed, since there is no further statement in that block to separate. However, the implied empty statement usually will do no harm: most compilers detect it and do not generate any null-operation code for that empty statement unless necessary.

Spaces and returns may appear anywhere in a program. Thus declarations, statements, *and comments* may occupy several lines if desired. Further, statements need not start in any particular column and can go on for as many lines as needed. This allows the careful programmer to lay out his program with indentation and multiple line usage to make it more readable.

No compound symbol can be separated by spaces. These include:

```
:=      assignment
(*      comment start
*)      comment end
..      variable range
<=      less than or equal
```

< >	not equal
> =	greater than or equal

In addition, no Pascal instruction (called a *reserved word* or *verb*) or variable name may contain spaces. Thus we can write

```
READ(A,B);
```

but not

```
RE AD(A,B);
```

ARITHMETIC OPERATIONS

We have already introduced a + sign as an arithmetic operator, and here we summarize the other arithmetic symbols:

+	addition
−	subtraction
*	multiplication
/	division with real number result
DIV	division with integer result
MOD	remainder after division of integers

The quotient in integer division is defined as the integer less than or equal to the quotient. The remainder is ignored. Thus

7/3 becomes 2.3333

but

7 DIV 3 becomes 2

Likewise, the operation A MOD B means the remainder after dividing *a* by *b,* so that

7 MOD 3 is 1

In complex expressions, multiplication and division are performed left to right before addition and subtraction. To be sure that an expression is evaluated as you expect, the use of parentheses is encouraged. For example,

```
5 + 3 * 6 = 23
```

but

```
(5 + 3) * 6 = 48
```

Only parentheses may be used, square brackets and braces are not permitted.

ARITHMETIC FUNCTIONS

The following arithmetic functions are also provided in Pascal and operate on the variable or expression inside the parentheses:

ABS(X)	absolute value of X
ARCTAN(X)	arctangent of X (radians)
COS(X)	cosine of X (radians)
EXP(X)	e to the power of X
LN(X)	ln(X)
LOG(X)	log(X) (not available in all versions)
ROUND(X)	round real to integer
SIN(X)	sine of X (radians)
SQR(X)	X*X
SQRT(X)	square root of X
TRUNC(X)	truncate real to lower integer

Note that not all possible functions are represented here. In fact the addition of LOG is an extension to standard Pascal, since it can always be obtained by dividing LN(X) by 2.303. Similarly, there is no function for raising an arbitrary number to an arbitrary power. Since such numbers are always calculated through the use of logarithms, the designers of Pascal decided to leave this function for the user to program by simply taking the logs and reexponentiating using the EXP function. For example,

$2^{14.3}$ is calculated by:
TWO14:= EXP(14.3*LN(2));

A REVISED PROGRAM FOR ADDING NUMBERS

In this program we will add together two numbers that are entered in the file INPUT with the result written to the file OUTPUT. In minicomputer and DEC-10 implementations of Pascal, these input and output files are actually the terminal keyboard and printer or display.

```
PROGRAM SUMMER;
VAR
     A,B,C:REAL;                (*DECLARE VARIABLES*)
BEGIN                           (*PROGRAM STARTS HERE*)
     READLN(A,B);               (*READ FROM INPUT FILE*)
     C:=A+B;                    (*PERFORM SUMMATION*)
     WRITELN('C = ',C);         (*WRITE RESULT INTO OUTPUT*)
END.                            (*THEN STOP AND EXIT*)
```

This program has some marginal use, since it actually performs some calculations on entered data. The data read in can be varied and the result inspected at the terminal. In a similar fashion, we could calculate the quadratic formula:

X:= (−B+ SQRT(SQR(B)−4*A*C))/(2*A);

or any other mathematical function. Note, of course, that we cannot calculate both roots in the same expression but would have to write:

X:= (−B− SQRT(SQR(B)−4*A*C))/(2*A);

for the other root.

The great power of Pascal or any other computer language lies in its ability to handle various types of data and make decisions based on intermediate results. These features are discussed in Chapters 4 and 5.

THE PYTHAGOREAN THEOREM

Now let us suppose we are going to write a simple, conversational Pascal program from scratch. The program will apply the Pythagorean theorem to calculate the length of the hypotenuse of a right triangle using the formula:

$$c = (a^2 + b^2)^{1/2}$$

We will take advantage of the long variable names in Pascal and name our variables SIDE1, SIDE2, and HYPOTENUSE. Since they could take on virtually any values, we will declare them as real:

```
VAR
    SIDE1, SIDE2, HYPOTENUSE:REAL;
```

```
PROGRAM PYTHAGORAS;
(*THIS PROGRAM CALCULATES THE LENGTH OF THE HYPOTENUSE
OF A RIGHT TRIANGLE, GIVEN THE LENGTHS OF THE TWO SIDES,
USING THE PYTHAGOREAN THEOREM*)
```

The only other important feature we will include is the *prompt message,* which simply sends a message to the user, telling him what data is expected:

```
ENTER THE LENGTHS OF THE SIDES:
```

Using the SQR (square) and SQRT (square root) functions, we see that the mathematical expression to be evaluated is:

```
HYPOTENUSE:= SQRT(SQR(SIDE1) + SQR(SIDE2));
```

The complete program is shown below:

```
PROGRAM PYTHAGORAS;
(*THIS PROGRAM CALCULATES THE LENGTH OF THE HYPOTENUSE
OF A RIGHT TRIANGLE, GIVEN THE LENGTHS OF THE TWO SIDES,
USING THE PYTHAGOREAN THEOREM*)

VAR
    SIDE1, SIDE2, HYPOTENUSE: REAL;

BEGIN
    WRITE('ENTER THE LENGTHS OF THE TWO SIDES: ');
    READ(SIDE1,SIDE2);

(*CALCULATE THE HYPOTENUSE USING THE THEOREM*)
    HYPOTENUSE := SQRT(SQR(SIDE1) + SQR(SIDE2));
    WRITELN( 'HYPOTENUSE= ', HYPOTENUSE);
END.
```

PROBLEMS

1 How would Pascal evaluate the following if X is real?

```
X:=5+4*6/7;
```

2 What value will Y have in

```
Y:=TRUNC(LN(5.0E7));
```

3 Which of the following are illegal? Why?

```
BEGIN;
C:=A-B+C;
5:=SQR(2.2);
(* COMMENT! * )
WRITELN(*A,B*);
```

CHAPTER 3

Basic Data Types in Pascal

We have already seen that we must declare both our constants and our variables before using them in Pascal programs. Further, since the types of values which the variable may represent are not discernible in the way constants are, we must declare each group of variables as having a certain type. In Chapter 2 we used only real numbers, and indeed for most scientific calculations, these will be the primary data type. However, Pascal permits several other standard data types, as well as having the ability to define new types of data.

THE REAL TYPE

Real numbers are numbers that can vary over an exceedingly large range from very small to very large, as well as having positive or negative signs. Real numbers can be represented by digits on both sides of a decimal point in standard fractional notation:

```
5.45
```

or in exponential notation, which is the computer equivalent of scientific notation:

```
4.32E5
```

means 4.32×10^5 (*not* e^5 !), and

```
6E7
```

means 60,000,000.

Real numbers in medium- to large-scale computer systems may occupy only one computer word, but in minicomputers and microcomputers, at least 2 words (or 4 bytes) are required to represent the sign, the number, and its exponent. We need not concern ourselves with exactly how a given computer represents a real number internally other than to note that real numbers usually occupy two words of storage in minicomputers. Thus if we make the declaration

```
VAR A,B,C: REAL;
```

we may be telling the compiler to reserve two words of storage for each of these three variables. In most computer systems real numbers may at least take on any value from 10^{-35} to 10^{35} and may be positive, negative, or zero.

THE INTEGER TYPE

Variables or constants having the *integer* type may only be whole numbers and can only be represented in a smaller range, usually from -2^{w-1} to $2^{w-1}-1$, where w is the worldlength of that computer. Integer calculations require only one computer word per integer and usually occur much faster than real calculations. Further, data acquired into the computer from data acquisition devices are nearly always in integer form, one word per data point. In the usual 16-bit minicomputer, as well as in most Pascal systems implemented through p-code interpreters, the integer range lies between $-32,768$ and $+32,767$. The actual value of the largest positive integer is contained in the predefined constant MAXINT, which may be referred to in your programs without declaring it.

Integers are automatically converted to real numbers within an arithmetic expression, when both real and integer variables are used, and such a result is always real. Real numbers are only converted to integers by the ROUND-(real) and TRUNC(real) functions.

THE BOOLEAN TYPE

Boolean variables can take on only the values true and false, and Boolean constants can only be set to TRUE and FALSE. They are usually used as markers or "flags" to indicate whether a particular condition may exist or not:

```
QUADETECT:=TRUE;
DATASWAP:=FALSE;
```

Boolean variables can also be set from the evaluation of an expression:

```
ISDONE:= SCANS=100; (*TRUE ONLY IF SCANS = 100*)
```

This expression sets ISDONE equal to TRUE if and only if SCANS=100, and FALSE for all other values of SCANS. More complex expressions are also possible:

```
NEUTRAL:=(6.9 < PH) AND (PH < 7.1);
```

The Boolean variable NEUTRAL will be set to TRUE only if PH lies between 6.9 and 7.1.

The operators that can operate on Boolean variables are:

```
AND
OR
NOT
```

The AND operator provides a TRUE result if and only if both operands are TRUE. If B1 is TRUE and B2 is TRUE, then (B1 AND B2) = TRUE. The OR operator provides a TRUE result if *either* operand is TRUE. If B1 is FALSE and B2 is TRUE, then (B1 OR B2) = TRUE, but if B1 and B2 are FALSE, then (B1 OR B2) = FALSE. The NOT operator reverses the state of any Boolean variable. If B1 is TRUE and B2 is FALSE, then (NOT B1) = FALSE and (NOT B2) = TRUE.

The truth tables for the AND and OR are shown below:

	B2	
AND	F	T
F	F	F
T	F	T

		B2	
OR		F	T
B1	F	F	T
	T	T	T

The symbols that can be used for comparing values are:

<	less than
<=	less than or equal to
=	equal to
< >	not equal
>=	greater than or equal to
>	greater than

An expression comparing two values has a Boolean value. Thus we can say:

```
B1 := R<=5.6;
```

and B1 will be set to TRUE if and only if R <= 5.6.

THE CHAR TYPE

Constants and variables having the type CHAR are variables that represent a single, printable character. The space is a printing character, but carriage returns, line feeds, tabs, and other such characters are not. A character may be represented within a program by enclosing it in apostrophes (or single quotes):

```
LETTERA:= 'A';
SPACE:=' ';
```

The apostrophe itself is represented by two apostrophes in a row:

```
APOS:=''''; (*APOSTROPHE CHARACTER*)
```

which are themselves enclosed in apostrophes as before.

Characters may be compared, but outside of the alphabet and the integers 0 to 9, the results will be dependent on what encoding scheme is used to represent characters in a particular computer system. Thus

'J' < 'K' is always

TRUE

and

'4' > '3' is always TRUE

but the value of

'1' < 'B'

will depend on the computer system.

Note also that there is a difference between a number and the character used to print that number, such as the *number* 3 and the *character* 3. The first will always have the absolute value 3, while the character 3 will have some numerical representation that will vary from computer to computer but will seldom be exactly equal to the numeric value 3.

Characters can be compared, but they cannot have any useful arithmetic operations performed on them directly. The usual application of the CHAR type is to get a one-letter command from the keyboard or an input file that instructs the program what to do next. Characters may also be used in text manipulation programs.

There are two special functions that operate in conjunction with character variables. If the variable C is of type CHAR and X is of type INTEGER;

```
X:=ORD(C)
```

puts the numerical value that represents the character variable C into variable location X. X must be of type INTEGER, and C must be of type CHAR.

```
C:=CHR(X)
```

puts the character represented by integer X into CHAR variable C.

These are actually only operations as far as the compiler is concerned, since the variables C and X have the same *numerical* value. The operations are used to tell the compiler that you know you are assigning the value of one type of variable to another type of variable.

While the actual numerical value of a character will vary from computer to computer, it is always true that

```
ORD('3')-ORD('0') = 3
```

or, in general, that

```
ORD('n')-ORD('0') = n
```

and that

```
CHR(ORD(n)) = n
```

where n is an integer.

Note that CHR(n) is only defined for integers that actually represent character codes. Other values will cause an error.

The functions PRED(ARG) and SUCC(ARG) operate on variables of the type CHAR (as well as on integers). The function SUCC(CH) produces the next character alphabetically after character CH, and the function PRED(CH) returns the character preceding character CH. Thus if

```
C1:='B';
```

then

```
C2:=SUCC(C1);      will set C2 to 'C'
```

and

```
C3:=PRED(C1);      will set C3 to 'A'
```

The values of PRED('A') and SUCC('Z') may not be defined, and according to the proposed standard, these operations should cause an error if the character is not defined.

THE STRING TYPE

A series of characters that can be manipulated as a single entity is called a *string.* The string may contain any of the printing characters, and when used as a constant within a program, it must be enclosed in apostrophes. The apostrophe character itself, if present, must appear twice in succession as in the CHAR variable case. Thus we can have:

```
WRITELN('DON''T YOU THINK THIS WILL WORK?');
```

but we cannot write:

```
WRITELN('THIS WON'T WORK');
```

In standard Pascal, such as is available on the DEC-10, strings can be compared only if they are the same length and are equivalent in type to a PACKED ARRAY OF CHAR (see Chapter 10). Strings can be entered at the keyboard or from an input file, but they must be made the same length to be compared with a constant string within a program. String operations are used primarily to interpret commands for a program to perform a number of operations, and have little other value except for alphabetizing lists and other elementary business applications.

In UCSD Pascal, strings may take on variable lengths, depending on how many characters are actually entered. Such strings are said to have a *dynamic* length. In such compilers, comparison of strings of different lengths is allowed, and the comparison is alphabetical, such that

```
'ABACUS' < 'ZEBRA'
```

even though 'ZEBRA' has less characters in it. Strings must always be less than one terminal line long since they can only be read in using the READLN statement, where the carriage return is used to terminate the string.

During output, strings entered in WRITE statements may not be continued onto the next line and should be broken into two or more WRITE statements when necessary.

```
WRITE('FOURSCORE AND SEVEN YEARS ');
WRITE('AGO, OUR FATHERS . . .');
```

will thus print

```
FOURSCORE AND SEVEN YEARS AGO, OUR FATHERS . . .
```

We can also have constant strings:

```
CONST
   FIRSTRING='FOURSCORE AND SEVEN YEARS ';
   LASTRING='AGO, OUR FATHERS . . .';
```

and print them out;

```
WRITELN(FIRSTRING,LASTRING);
```

PROBLEMS

1 Write a Pascal expression setting X = 5 raised to the 11th power.

2 Which of the following Pascal expressions are illegal? Rewrite them correctly.

```
VAR X,Y:REAL
INTG1,INTG2,M,N,:INTEGER;
CK:CHAR;
```

```
B,B1:BOOLEAN;
BEGIN
    Y=X+Y;
    J:=INTG1+N;
    X:=M*Y;
    M:=X*Y;
    Y:=M*X+B↑2;
    Y:=Y+13.6;
    CK:=M+N;
    B:=FALSE;
    B1:=B;
    X:=Y+CK;
    B1:=X<Y;
END;
```

3 How could the long string in the previous section be printed out using only one WRITE statement?

4 Write a Pascal statement setting B to TRUE if M is not equal to N and FALSE otherwise.

5 Write a program to calculate the volume of a sphere from the entered radius, and print it out.

6 A Lorentzian line is given by

$$g(\nu)=\frac{g_{max}}{1+4\pi^2T_2^2(\nu-\nu_0)^2}$$

where $\pi=3.1416$, $T_2=1/(\pi\times \text{linewidth})$, and $(\nu-\nu_0)$ is the distance from the center of the peak ν_0, where the height is g_{max}. Write a Pascal expression for $g(\nu)$.

CHAPTER 4

Declarations in Pascal

THE PROGRAM HEADER

A program may begin with the word PROGRAM followed by a name. This statement is required by some compilers, optional in others, and not allowed in a few more. If the PROGRAM header occurs, it must be the first noncomment statement in the program. The program name may have any number of characters, but usually only the first 6 to 8 are significant and are passed on to the linking loader.

CONSTANT DECLARATIONS

We have already seen the use of the CONST declaration in Chapter 2. However, we shall detail its syntax here. Constants may be used within the program without declaring them in advance:

```
CONST
   TEMP=273.16;   (*REAL CONSTANT*)
```

However, it is a principal tenet of the well-organized structured program that all constants be located in one place at the beginning, so that they can be changed easily if the conditions for using the program change. For example, on another day it might be sensible to make TEMP equal to 289.0, and we would otherwise have to go looking through the program for this and all related constants to change them. Referring to constants by name eliminates this problem and allows us to make only one change in the program.

In most programs the CONST declaration will come first, right after the PROGRAM declaration and any introductory comments. It must come *before* the VAR declaration section. The format is:

```
CONST
   NAME1=VALUE1;
   NAME2=VALUE2;
```

and so on.

The names must be on the left side, followed by an *equal* sign (note: *not* an := pair), and terminated by a semicolon. The type of the constant is determined by the compiler from the format.

1 REAL constants must have digits on both sides of the decimal point or the E.

2 INTEGER constants must have no fractional or exponent part, or decimal point.

3 CHAR constants must be a single character enclosed in apostrophes.

4 BOOLEAN constants must be set to either TRUE or FALSE.

5 STRING constants must contain two or more characters enclosed in apostrophes.

THE VAR DECLARATION

The VAR declaration section must follow the CONST section immediately and occur before any executable program statements. Following the VAR declaration must come groups of one or more variables by name belonging to that type. Each in a series must be separated by commas and the last in the series followed by a colon, followed by the type name. The type name must be followed by a semicolon. Additional variables of a different type (or more of the same type in another list) may follow in the same fashion until all variables of all types have been declared:

```
VAR
   A,B,CRUD:REAL;
   XWORD,SCANS:INTEGER;
   QUADETEC,DATASWAP:BOOLEAN;
```

While it may seem troublesome to declare variable names in advance, this is an important component of the well-planned structured program, which allows the compiler to check for variable misspellings and inconsistencies of type.

NAMES OF VARIABLES AND CONSTANTS

Pascal allows variable or constant names to have any length desired. They must begin with a letter. Following the first letter, any combination of numbers and letters is allowed. Grogono (1) has noted that it is bad practice to use the letter O and the digit 0 in any way that might be confused in a variable name. Likewise, the tail of the letter Q may not always be clear and may thus be confused with O or 0. Lowercase l can also be confused with the numeral 1.

While variable names may have any practical length, the original Pascal report noted that it is acceptable to compare only the first *eight* characters of such a name in determining whether a variable name is unique. This is the approach used in most minicomputer and microcomputer implementations of Pascal, and we will use it here. Decsystem-10 Pascal actually differentiates between variable names up to 10 characters long.

Thus you should, in general, be careful not to choose variable names that differ only in characters after the eighth. For example, the compiler would regard CHLOROFORM and CHLOROFORMD as identical variable names. This can usually be solved by choosing a new set of names such as CHCL3 and CDCL3. In any case, the best choice of a variable name is one that reminds you of *what it represents.* Thus we might avoid using A, B, or X, and instead use OPPOSITE, ADJACENT and HYPOTENUSE.

Some Pascal compilers allow the use of the underscore character (_) to separate parts of longer variable names to improve readability. Instead of using the run-together name SUMOFPAIRS, you could then use the name SUM_OF_PAIRS. Since the underline characters are often ignored, they do not count in the number of unique characters required in a variable name. Variable names can also be made more readable by mixing uppercase and lowercase when the terminal and printer allow this: SumOfPairs.

BEGIN AND END

The declarations BEGIN and END are sometimes confusing to the programmer new to Pascal because they do not seem to be statements causing any action, nor are they declarations telling the compiler what to do. These declarations are best regarded as flags to the compiler, telling it where *blocks* of code begin and end.

A block of code is simply two or more statements that will always be executed together. Any place that a statement can be placed, it is legal to place a block of statements set off by BEGIN and END. It is this feature that makes Pascal so useful in writing well-structured programs. To appreciate this block structure, we shall next take up the decision-making features of Pascal.

PROBLEMS

1 What type is each of the following constants?

```
CONST
   DPOS='''':
   PI=3.1416;
   FOUR=4;
   NAME='NAUGHTYBITS':
   A='A';
   NINE=9.0;
```

2 Which of the following are illegal variable names? Which are unwise?

```
AARDVARKS
RUT_A_BAGA_S
CONST
5FOLD
BEGINS
'SMILE'
IN1T
TO0
```

REFERENCES

1 P. Grogono, "Programming in Pascal," revised edition, Addison-Wesley, Reading, Mass., 1980.

2 J. Welsh and J. Elder, "Introduction to Pascal," Prentice-Hall International, Englewood Cliffs, N.J., 1979.

3 G. M. Schneider, S. W. Weingart, and D. M. Perlman, "An Introduction to Programming and Problem-Solving with Pascal," Wiley, New York, 1978.

CHAPTER 5

Decision-Making and Loop Control in Pascal

The great power of all computer languages and, indeed, of computers themselves is the power to make decisions during the course of a program. These decisions may be as simple as whether to continue an iteration process or as complex as a multibranched pathway depending on a number of possible variables. In Pascal the decision-making instructions are especially powerful and highly *structured,* which means that their form is easy to read and that they are easy to write in an unambiguous way. One of the features of Pascal which makes this possible is the free format which allows statements to be indented by as many spaces as desired. We will show the indentation in these statements, and we encourage you to use this form to make your programs more easily debuggable and understandable, both to others and to yourself a month after you have written them.

THE IF-THEN STATEMENT

The simplest form of decision making in Pascal is the IF-THEN statement. It has the general form:

IF Boolean expression THEN statement:

and means that if the Boolean expression is TRUE, then the statement following the THEN will be executed. A typical example from everyday life might be the instructions to a friend:

"If they have chocolate ice cream, bring me an ice-cream cone."

We might phrase this in Pascal as:

```
IF ICECREAM = CHOCOLATE
THEN CONE:= TRUE;
```

Note that there are no semicolons in the IF statement. In fact, placing a semicolon after the THEN will imply an "empty statement" as a result of the THEN and will cause an entirely different action than expected. For example,

```
IF ICECREAM = CHOCOLATE THEN;    (*UNWANTED SEMICOLON*)
CONE:=TRUE;             (*ALWAYS EXECUTED*)
```

will execute the "empty" statement after the THEN if ICECREAM is CHOCOLATE, but will execute the following statement CONE:=TRUE whether the ice cream is chocolate, vanilla, or tutti-frutti. Some Pascal compilers flag this semicolon as a syntax error.

Let us now consider a simple program to detect whether a number is odd or even. In this first example, we will print out the message NUMBER IS EVEN if it is, in fact, even. The method we will use to detect even integers relies on the fact that the DIV operation produces an integer quotient and no fractional part.

So the value of

5 DIV 2   is 2

while the value of

6 DIV 2   is 3

Thus to detect an even number, we need only see if that number divided by 2 and then multiplied by 2 is unchanged. Thus

(6 DIV 2)*2   is still 6

but

(5 DIV 2)*2   is only 4

The program below asks for a number at the terminal and prints out the result only if the result is even. If the number is odd, it simply exits without printing any message.

```
PROGRAM IFDEMO1;
(*THIS PROGRAM READS IN INTEGERS AND
TELLS THE USER WHETHER THEY ARE EVEN*)

VAR NUM:INTEGER;

BEGIN                    (*PROGRAM STARTS HERE*)
    WRITE('ENTER AN INTEGER: ');
    READLN(NUM);

    IF((NUM DIV 2)*2 = NUM)
        THEN WRITELN('NUMBER ',NUM,' IS EVEN');
END.
```

Note again, that the commas separating the optional output file name OUTPUT, the messages, and the variable names are *required.*

This program accepts an integer input from the terminal, and if and only if it is even, does it print out the message NUMBER NNN IS EVEN. It then exits from the program in any case. There is actually an ODD function in Pascal which returns the value TRUE if an integer is odd, but we have just generated it ourselves. We would use the ODD function as follows:

```
IF NOT ODD (NUM) THEN
  WRITELN('NUMBER', NUM, 'IS EVEN');
```

MULTIPLE COMPARISONS

An IF clause can be made up of several Boolean expressions combined together to form a single result. Each Boolean operation, however, can consist of only two expressions, one on each side of the Boolean operator. For example, we can write:

```
IF (6.9<PH) AND (PH<7.1) THEN statement
```

but we cannot write:

```
IF (6.9 < PH < 7.1) (*THIS IS WRONG!!*)
```

since it is not clear what operations are being performed.

The IF-THEN-ELSE STATEMENT

A somewhat more versatile form of the IF statement has the form:

```
IF Boolean expression
   THEN statement1
   ELSE statement2;
```

This powerful statement will cause the execution of statement1 if the expression is TRUE and then will *skip* statement2. If the Boolean expression is FALSE, the program will skip statement1 and *execute* statement2. To illustrate this feature, we simply expand our even number detector to print out a message if the number is odd as well. Note that only one of the two cases can ever be executed.

```
PROGRAM IFDEMO2;
(*THIS PROGRAM READS IN INTEGERS AND
TELLS THE USER WHETHER THEY ARE ODD OR EVEN*)

VAR NUM:INTEGER;

BEGIN                    (*PROGRAM STARTS HERE*)
     WRITE('ENTER AN INTEGER: ');
     READLN(NUM);

     IF((NUM DIV 2)*2 = NUM)
          THEN WRITELN('NUMBER ',NUM,' IS EVEN')
          ELSE WRITELN('NUMBER ',NUM,' IS ODD');
END.
```

Note further that there are *no semicolons* anywhere in the IF-THEN-ELSE statement, and inserting one between the THEN and ELSE clause will cause the compiler to generate an error message.

COMPOUND STATEMENTS OR BLOCKS

As we have already noted, Pascal allows the replacement of any statement with a series of statements forming a block or compound statement, preceded by BEGIN and terminated by END. One of the most common places for the compound statement is in IF statements. Rather than just doing one thing when an IF is satisfied, a whole series of operations (including further subsidiary IF statements) may be performed.

In the program below, we show how to calculate the roots of a quadratic equation of the form:

$$ax^2 + bx + c = 0$$

If the expression under the square root is negative, we simply print "roots are imaginary." Note the Boolean expression following the IF, the compound statement in the THEN clause, and the simple statement in the ELSE clause.

COMPOUND IF STATEMENTS

Another IF statement can be used as the THEN or ELSE clause of an IF statement. This leads to syntax like the following:

```
IF Boolean1
   THEN statement1
   ELSE
      IF Boolean2
         THEN statement2
         ELSE statement3;
```

```
PROGRAM QUADFORM;

(*THIS PROGRAM CALCULATES THE REAL
ROOTS OF THE QUADRATIC EQUATION AX^2 + BX + C =0.
IF THE ROOTS ARE IMAGINARY, THE MESSAGE
"ROOTS ARE IMAGINARY" IS PRINTED. *)

VAR
     A,B,C,X1,X2,R :REAL;

BEGIN
     (*GET A,B AND C FROM THE TERMINAL*)
READ(A,B,C);

(* CALCULATE THE TERM UNDER THE SQUARE ROOT*)

R:= SQR(B)-4*A*C;

(*SEE IF THE ROOTS ARE REAL OR IMAGINARY*)
IF R>=0.0
THEN
     BEGIN
     X1:=(-B+SQRT(R))/(2*A);
     X2:=(-B-SQRT(R))/(2*A);
     WRITELN('ROOTS ARE',X1,X2);
     END
ELSE
     WRITELN('ROOTS ARE IMAGINARY');
END.
```

These sorts of constructions are perfectly legal and sometimes useful, but are quite confusing to read and should only be used when one of the other constructions, notably the CASE statement, cannot be used.

THE "DANGLING ELSE" CONSTRUCTION

One further reason for avoiding compound IF statements is the potential confusion that can arise when some of the IF statements have ELSE clauses and others do not. It is sometimes difficult to be sure what the program means in such cases. Consider:

```
IF b1 THEN IF b2 THEN s1 ELSE s2;
```

This can be made somewhat clearer by indenting:

```
IF b1
   THEN IF b2
      THEN s1
      ELSE s2;
```

The rule is that the ELSE statement belongs to the closest IF that has no ELSE clause. Since this can be so confusing, it should be avoided.

THE REPEAT-UNTIL STATEMENT

One of the things that computer programs do best is repetition of the same instructions. This can be a true mathematical iteration process, or it can be a simple input loop. One common looping control statement is the REPEAT-UNTIL statement which has the form:

```
REPEAT
   statement(s)
UNTIL Boolean expression;
```

This statement or the sequnce of statements separated by semicolons between the REPEAT and the UNTIL are repeated until the Boolean expression becomes TRUE. Note that the REPEAT-UNTIL may enclose a *block* of statements *without* the use of BEGIN and END. A typical example of this for the control of input data is provided in our revised program for testing odd or even numbers. In this version new numbers are continually requested until a zero is entered. Then, *after* the calculation for this last value has been performed, the program ends. Note that the test for the Boolean expression is done at the *end* of the loop.

```
PROGRAM IFDEMO3;
(*THIS PROGRAM READS IN INTEGERS AND
TELLS THE USER WHETHER THEY ARE ODD OR EVEN*)

VAR NUM:INTEGER;

BEGIN                    (*PROGRAM STARTS HERE*)
REPEAT                   (*REPEAT LOOP STARTS HERE*)
    WRITE('ENTER AN INTEGER: ');
    READLN(NUM);         (*READ IN A VALUE*)

    IF((NUM DIV 2)*2 = NUM)
        THEN WRITELN('NUMBER ',NUM,' IS EVEN')
        ELSE WRITELN('NUMBER ',NUM,' IS ODD');
UNTIL NUM = 0;   (*EXIT IF NUMBER IS 0*)
END.
```

THE WHILE-DO STATEMENT

A second form of loop control is available in Pascal in the WHILE-DO statement. This statement has the form:

WHILE Boolean expression
  DO statement;

It has the meaning that the statement is executed while the Boolean expression is TRUE. Of course, the statement may be a compound one. Note that the WHILE-DO statement checks the expression at the *top* of the loop, *before* executing any statements, while the REPEAT statement checks the expression *after* executing all statements in the loop. Note that, unlike REPEAT, the WHILE-DO construction operates on only *one* statement, and if several are to be used, they must be enclosed in a BEGIN-END pair.

One reason for using the WHILE rather than the REPEAT statement might be that some part of the expression to be evaluated within the loop will become undefined (such as when division by zero occurs) if a certain variable becomes too small. This is illustrated in the program below, which calculates factorials of numbers:

```
PROGRAM FACTOR;
(*CALCULATES THE FACTORIAL OF ANY ENTERED INTEGER*)
VAR
     N,NSAVE:INTEGER;
     FACTORIAL:REAL;

BEGIN
REPEAT                        (*GET EACH NUMBER AT TERMINAL*)
     WRITE('N= ');            (*PROMPT USER*)
     READLN(N);
     NSAVE:=N;                (*SAVE N FOR CHECK AT END*)
     FACTORIAL:=1;            (*START AT 1*)

     (*CALCULATE FACTORIAL UNTIL N IS REDUCED TO 0*)
     WHILE N>0 DO                (*CHECK N BEFORE ENTERING LOOP*)
          BEGIN
          FACTORIAL:=FACTORIAL*N;
          N:=N-1;                (*REDUCE BY 1 EACH TIME*)
          END;

     (*WRITE OUT FACTORIAL IF N WAS >0 *)
     IF NSAVE >= 0 THEN
          WRITELN(NSAVE, ' FACTORIAL = ',FACTORIAL);
UNTIL NSAVE<0;                   (*QUIT IF N<0 *)
END.
```

ARRAYS

In order to discuss useful applications of some of the other looping control statements, we will introduce the concept of arrays here. An *array* is simply a list of values having the same name and type but different subscripts. In mathematics we might speak of

x_i where i is the index of a set of x's, say $x_1,.., x_{20}$

In Pascal we do the same, except that we enclose the subscripts in *square brackets* since most terminals do not have subscripting capabilities. Thus the array here becomes X[I] and a particular element, say, X[19].

As with all other variables, arrays must be declared in advance by both type and size. Arrays may be REAL, INTEGER, BOOLEAN, or CHAR, as well as some more complex types defined within the program. The array must be declared in the declaration section:

```
name: ARRAY[m..n] OF type;
```

where m and n are the lower and upper indices of the array (it need not start at 1!), and *type* is one of the standard types or a specially defined type. For example, we might define an array of 20 real variables as:

```
X: ARRAY[1..20] OF REAL;
```

Note that square brackets are used in the definition and to enclose the actual subscripts. Parentheses are used in Pascal only in function calls, in READ and WRITE statements, and to control the order of evaluation of arithmetic expressions. Multidimensional arrays are declared analogously, using a comma to separate the dimensions:

```
X2: ARRAY[1..20,1..30] OF REAL;
```

although more efficient declarations of such arrays are discussed in Chapter 10. Individual elements are then referred to as X[I,J] or as X[15,17], for example.

It is important to note that the actual size of any array must be declared in the VAR declaration section using constants. Neither of the limits may in itself be a variable. The constants may be named constants or numeric constants, however:

```
CONST
   LOW=2;  HIGH=27;
VAR
   X:ARRAY[LOW..HIGH] OF REAL;
```

Array indices must be of type INTEGER, but may be positive, negative, or zero if this is useful to you.

THE FOR-DO STATEMENT

The FOR statement is used to control repetition of a statement or block when the number of repetitions can be calculated in advance. The form of the FOR statement is:

```
FOR loopvar:= lowerlimit TO upperlimit DO
   statement;
```

The statement can, of course, be compound, and the lower and upper limits can be of type INTEGER or CHAR, but not REAL, and may be expressions involving integer or character variables. Specifically, we might write:

```
FOR I:= 1 TO 20 DO
   SUM:= SUM + X[I];
```

Note that unlike REPEAT-UNTIL, the FOR-DO (and the WHILE-DO) operate only on *one* statement, and a BEGIN-END pair must be used to enclose a block of statements to be passed through.

A second form of the FOR statement allows you to go through a loop while varying a number from a high value to a lower value, thus allowing you to go through an array backwards, if you wish. This has the form:

```
FOR loopvar:= uppervalue DOWNTO lowervalue DO
   statement;
```

and might specifically be written as:

```
FOR I:= 20 DOWNTO 1 DO
    SUM:= SUM+X[I];
```

In both cases note particularly that the FOR variable, called the *index* variable, can only be varied in steps of 1. It is also important to note that if the lower limit is greater than the upper limit, the FOR loop will not be executed *at all!* (This is unlike FORTRAN where the loop is always executed once.)

These two forms as well as the declaration of arrays are illustrated in the program NORMAL given below, which accepts up to 20 numbers from the terminal, finds the largest one entered, and divides through by it.

Note that the first loop is a REPEAT loop which allows entry of the X values. There are two cases under which repetition of the loop is to terminate:

```
PROGRAM NORMAL;

(*READS IN UP TO 20 NUMBERS AND THEN
FINDS THE LARGEST OF THE SERIES
AND DIVIDES THROUGH BY IT *)

CONST
    TOP=20; (*INTEGER MAXIMUM OF ARRAY*)

VAR
    N,I: INTEGER;
    X: ARRAY[1..TOP] OF REAL;
    XMAX:REAL;

BEGIN
I:=0;
REPEAT
(*START BY SETTING I TO 1 AND READING NUMBERS
UNTIL ONE <= 0 IS FOUND*)
    I:=I+1;        (*ADVANCE COUNTER*)
    READ(X[I]);
UNTIL (X[I]<= 0.0) OR (I>=TOP);
(*QUIT IF ARRAY IS FULL OR NUMBER<=0 IS ENTERED*)

IF X[I]=0.0 THEN N:=I-1 ELSE N:=I;


XMAX:=0.0;   (*THIS WILL BE THE LARGEST FOUND*)

FOR I:=1 TO N DO
    IF XMAX < X[I] THEN XMAX:=X[I];

(*HAVING FOUND XMAX, DIVIDE THROUGH BY IT*)
IF XMAX >0.0 THEN     (*THIS MAKES SURE SOME NUMBERS
                        WERE ENTERED*)
    FOR I:= N DOWNTO 1 DO
        BEGIN
        X[I]:=X[I]/XMAX;
        (*WRITE OUT RESULTS*)
        WRITELN('X[', I:2, ']=', X[I]);
        END;
END.
```

1 If a zero value is entered, or

2 If more than 20 have been entered

This leads to a Boolean condition made up of two simpler conditions:

```
(X[I] <= 0.0)     or     (I >= TOP)
```

They are simply combined with the Boolean operator OR, which has exactly the meaning in Pascal that it appears to have from reading the statement: Repeat until X[I] = 0.0 *or* I >= TOP.

In the first FOR statement we see that the statement to be executed is in itself conditional: it is an IF-THEN statement. I is varied from 1 to N, and each time XMAX is checked to see if it is less than X[I]. If it is, XMAX is set equal to that value of X[I]. When the program has gone all the way through the list, XMAX will be equal to the largest value entered in the X array.

The second use of the FOR statement is in a loop where all the X's are divided through by XMAX. As an illustration of the DOWNTO feature, the array is traversed in reverse direction. This is simply for illustration, however; it could just as easily have occurred in the forward direction. So that you can see the result of normalization, the program prints out the normalized value of each X along with the message X[I]=, where I is the variable inserted in the printout. Look carefully at the use of apostrophes in the last WRITE statement, and note the use of the CONST declaration for the array bound 20.

In the last statement of the program NORMAL the symbol 'I:2' means: use just two columns to print out the value of I. Formatting output is discussed in Chapter 7.

MOVEMENT OF ENTIRE DATA STRUCTURES IN PASCAL

Pascal has the unique feature of being able to move entire data structures (usually arrays) in a single statement. Thus the data in array X may be copied into array Y by the simple statement Y:=X; when the arrays have the same dimension and type. Thus we see that the tedious statement

```
FOR I:=1 TO TOP DO Y[I]:=X[I];
```

can be replaced by

```
Y:=X;
```

A simple program illustrating this feature is given below:

```
PROGRAM BLKTST;
(*THIS PROGRAM ILLUSTRATES MOVEMENT OF ENTIRE ARRAYS
WITH A SINGLE STATEMENT *)

VAR
X,Y:ARRAY[1..20] OF REAL;    (*TWO EQUAL SIZE ARRAYS*)
I:INTEGER;

BEGIN
(*SET ARRAY X EQUAL TO ITS INDEX*)
    FOR I:=1 TO 20 DO X[I]:=I;

(*MOVE WHOLE ARRAY X INTO Y IN ONE STATEMENT*)
    Y:=X;
END.
```

DIFFERENCES AMONG THE LOOP CONTROL STATEMENTS

The great variety of overlapping capabilities among the IF-THEN, REPEAT, WHILE, and FOR statements can be confusing at first. However, you should recognize that these are similar statements and are all provided to help you in writing the most efficient, structured, readable program possible.

The FOR statement is the most restrictive of these, since it can only be used to pass through a set of statements a definite number of times. This can actually be accomplished with any of the others in a somewhat awkward way. Consider:

```
SUM:=0.0;
FOR I:=1 TO 10 DO
    SUM:=SUM+I;
```

This could also be written as:

```
I:=1;
SUM:=0.0;
REPEAT
    SUM:=SUM+I;
    I:=I+1;
UNTIL I>10;
```

or as:

```
I:=1;
SUM:=0.0;
WHILE I<=10 DO
BEGIN
   SUM:=SUM+I;
   I:=I+1;
END;
```

In the FOR statement we see that a variable can be incremented or decremented only in steps of 1. Further, that loop variable *cannot be changed* within the loop. Thus in the example above we cannot change I within the FOR loop controlled by I. In addition, when we leave the FOR loop, the value of I becomes *undefined* and we cannot check and use it as we could in the other two cases, unless we reset it first. FOR loops will not be executed at all if the lower limit is already greater than the upper limit.

Finally, the proposed standard requires that the loop variable be defined in the current procedure rather than in the outer, main, program. Procedures and local and global variables are discussed in Chapter 6. The proposed standard also requires that the FOR loop variable be an "entire" variable. This means that it cannot be an element of an array or record (Chapter 11).

In REPEAT statements we can change any variable by any amount within the loop and can exit on nearly any condition or combination of conditions. The condition is tested at the end of the loop, and thus the loop is always executed at least once. Further, only the REPEAT-UNTIL statement can surround more than one statement without the BEGIN-END pair.

In WHILE statements we can change any variables by any amount within the loop, but the condition is tested at the beginning of the loop. It is thus possible that the loop might not be executed at all if the test fails the first time. Before entering a WHILE loop, you should be sure that the variables that the WHILE will test are defined. If they have not yet been set to any value within the program, the results of the WHILE test will depend on the contents of memory when the program starts. These contents, of course, may be random and can vary from time to time, leading to unexpected results. A compiler conforming to the proposed standard (few do on this point) should issue a compile-time or run-time error in such a case.

WHEN TO USE EACH LOOPING CONTROL STATEMENT

Clearly, since we can accomplish nearly the same things with these three statements, we should have some rules of thumb for choosing among the three looping control statements.

1 Use FOR when going through an array or other structure when the exact number of repetitions can be calculated in advance:

```
FOR K:=0 TO MAX DO
    SUM:=SUM+X[K];
```

2 Use WHILE when there is a possibility that the loop should never be executed; that is, when the Boolean condition might already be satisfied:

```
SUM:=0;
WHILE X>0.0 DO
    BEGIN
    SUM:=SUM+1/X;
    X:=X-0.1;
    END;
```

Be sure that the condition is defined before entering the WHILE loop. In the above case this means to be sure that X has a known value before entering the loop.

3 Use REPEAT when the loop should always be executed at least once before the Boolean condition is expected to become TRUE:

```
I:=1;
REPEAT
    ODDSUM:=ODDSUM+ X[I];
    I:=I+2;
UNTIL I> TOP;
```

USE OF LOCAL VARIABLES IN FOR-DO LOOPS

The proposed standard requires that the loop variable in FOR-DO loops be a local variable to that procedure rather than a global variable. Compilers which follow that standard issue an error message if global variables are used, since they are undefined at the end of a loop and may take on surprising values after a procedure using one of them is called.

COMPARING REAL NUMBERS

A common mistake made by some programmers is to assume that two real numbers will become exactly equal when they are arrived at by two different processes. The loop

```
Y:=0.0;
REPEAT
   Y:=Y+0.1;
UNTIL Y=10.0;
```

may never finish executing, because it is impossible to represent 0.1 exactly in binary, and thus the running sum Y may never be *exactly* 10.0. Instead, you should write:

```
REPEAT
   Y:=Y+0.1;
UNTIL Y >= 10.0;
```

to be sure that the program will eventually finish this loop.

If it is critical that a certain number returned from some series of calculations be compared to exact real numbers, it is usually safer to see to it that the numbers agree within a small tolerance, such as 1.0E−10, rather than hoping that they will be exactly equal. Thus you should write:

```
CONST TOLERANCE = 1.0E-10;
BEGIN
   :
   :
IF ABS(X-273.16) <= TOLERANCE THEN statement
```

rather than writing:

```
IF X = 273.16 THEN statement
```

PROGRAMS REQUIRING SEVERAL CHOICES

It is sometimes necessary to execute several different statements depending on one of several different values encountered in a calculation. When there are only two or three choices, the IF-THEN-ELSE statement may be sufficient, but when there are more than three, more complex constructions may be necessary. One possibility might be the use of IF statements within IF statements, where the ELSE clause of each IF statement is, in itself, another IF statement.

For example, in a student score-keeping program, you might want to enter a score for a quiz, an exam, an experiment, or a homework assignment. In the score-keeper program that follows, one of the arrays QUIZ, EXAM, EXPT, or HMWK is specified, and then the experiment, quiz, exam, or homework number is specified, followed by the score. The nested IF method of making these choices is illustrated in this program fragment. (Note that such elegances as which student, and the printout of the results are omitted for brevity.)

```
PROGRAM SCRKEEPR;
(*THIS PROGRAM ACCEPTS THE COMMAND CHARACTERS
Q,E,X OR H, FOLOWED BY THE
NUMBER OF THAT ARRAY AND THE SCORE.*)

VAR
     SCRTYP: CHAR;     (*CHARACTER SHOULD BE Q,E,X OR H*)
     NUMBER: INTEGER;
     SCORE:  REAL;
     QUIZ:   ARRAY[1..4] OF REAL; (*4 QUIZZES*)
     EXAM:   ARRAY[1..4] OF REAL; (*4 EXAMS*)
     EXPT:    ARRAY[1..8] OF REAL; (*8 EXPERIMENTS*)
     HMWK:   ARRAY[1..10] OF REAL; (*10 HOMEWORK*)

BEGIN
(*START BY GETTING THE SCORE TYPE,
THE NUMBER AND THE SCORE*)

READ(SCRTYP, NUMBER, SCORE);

(*CHECK TO SEE WHICH ARRAY TO USE AND ENTER RESULT*)

IF SCRTYP='Q'
     THEN QUIZ[NUMBER]:=SCORE

     ELSE IF SCRTYP='E' THEN
          EXAM[NUMBER]:=SCORE

     ELSE IF SCRTYP='X' THEN
          EXPT[NUMBER]:=SCORE

     ELSE IF SCRTYP='H'
          THEN HMWK[NUMBER]:=SCORE;
END.
```

THE CASE STATEMENT

While, with judicious spacing, the nested IF method is readable, the logic quickly becomes tortuous and hard to follow. Therefore Pascal also contains the powerful CASE statement, which allows as many choices as desired. It has the form:

```
CASE expression OF
    value1: statement 1;
    value2: statement 2;
    value3: statement 3;
        (*etc. *)
END:
```

The CASE statement evaluates the variable or expression following the word CASE and then executes the statement following that value of the expression. If we just wanted to print out the value of X, we could, in a simple-minded way, use the CASE statement to do it:

```
CASE X OF
    5: WRITELN('X=5');
    90: WRITELN('X=90');
      (*etc. *)
    END(*CASE*);
```

Note that the list of cases must terminate with an END statement. Since ENDs may appear in several contexts, it is conventional to explain what one is doing in complex programs with a comment after the END. If the operations to be performed in each CASE alternative are more than one statement long, they are surrounded by the BEGIN-END pair as usual.

Now let us look at the great simplification the CASE statement gives to our score-keeping program. In this second version the command characters Q, E, X, and H are checked by the CASE statement, and only the statement(s) following that value of the case are executed. All other cases in the list are skipped. Standard Pascal expects that all cases must be accounted for and that if a value is encountered for which there is no CASE selector statement, the result is undefined.

In UCSD and Aspect Pascal, the next statement after the case END

```
PROGRAM SCRKPR2;
VAR
     SCRTYP: CHAR;
     NUMBER: INTEGER;
     SCORE:  REAL;
     QUIZ:   ARRAY[1..4] OF REAL;
     EXAM:   ARRAY[1..4]OF REAL;
     EXPT:   ARRAY[1..8] OF REAL;
     HMWK:   ARRAY[1..10] OF REAL;

BEGIN
READ(SCRTYP,NUMBER,SCORE);

(*CHECK TO SEE WHICH ARRAY TO USE
AND ENTER SCORE*)

CASE SCRTYP OF
     'Q': QUIZ[NUMBER]:=SCORE;

     'E': EXAM[NUMBER]:=SCORE;

     'X': EXPT[NUMBER]:=SCORE;

     'H': HMWK[NUMBER]:=SCORE;

END; (*CASE*)
END.
```

statement is executed if there is no match to any of the cases. In DEC-10 Pascal (CMU version), a run-time error will occur if a value of the CASE selector variable occurs which is not included in any of the cases. This can be avoided by restricting the CASE statement as described in the next section.

The CASE statement has the further power of allowing several choices to select the same block of code, as shown in the fragment below:

```
CASE Y OF
    2,3,4:     statement1;
    5,7,9:     statement2;
    6,8,11,1:  statement3;
    12:        statement4;
END: (*CASE*)
```

The values preceding the colons in a CASE statement are called *case selectors.* They must be of type CHAR, INTEGER, or BOOLEAN, be constants known to the compiler, and be of the same type as the CASE selector expression. They cannot be variables, although they can be named constants. Obviously, each selector value can appear only once within a CASE statement.

RESTRICTING THE CASE STATEMENT SELECTOR VALUES

The CASE statement assumes that any value that a variable can take on will be in one of the cases. If there is only a small range of values upon which the CASE statement is to act, values outside that range can be eliminated by using variables that are *subrange* types, as discussed in Chapter 10, or by checking whether a particular variable is a member of a *set* of variables. Sets are also described in Chapter 10, and an example of restricting a CASE statement with the test for set membership is given.

If there are only a few values for which no action is to be taken, they can be listed as selectors of an empty statement:

```
IF (X>0) AND (X<11) THEN
(*RESTRICT VALUES TO 1..10*)
CASE X OF
   9,1,7: statement1;
   6,4,2: statement2;
   3,5,8,10:  ;  (*EMPTY STATEMENT*)
END; (*CASE*)
```

A few versions of Pascal, notably some available on the DEC-10, allow the OTHERS statement as one of the CASE selectors, so that any value of the CASE selector variable that is not covered by the preceding cases will cause execution of the statement(s) following the OTHERS: statement. For example,

```
CASE Y OF
   2,3,4:     statement1;
   5,7,9:     statement2;
   OTHERS: statement3;
END:(*CASE*)
```

This code says: execute statement1 if Y has the value 2, 3, or 4: execute statement2 if Y has the value 5, 7, or 9; and execute statement3 for all other values of Y.

PROBLEMS

1 The Fibonacci series has the form 1, 1, 2, 3, 5, . . ., where each element of the series is the sum of the previous two. This series occurs in nature

and has been found to describe flower petals and pine cones. Write a program to print out all elements of the series less than 50.

2 Write a program to read in alist of 10 numbers and print out the smallest nonzero number entered.

3 Write a program to convert entered temperatures from Fahrenheit to Celsius if the number and an 'F' are entered and from Celsius to Fahrenheit if the number and a 'C' are entered.

4 Write a program to accept a number and a character. If the character is 'L', print out the log of the number; if the character is 'E', print out the exponential of the number; if the character is 'S', print out the square root of the number. Cycle until a negative or zero value is entered.

5 Write a program to find the longest continuous block of zero data in an array of 4096 points. Print out the first and last indices of the block.

6 The program fragment

```
X:=12.8;
Y:=X*10;
WRITELN(Y);
```

produces an output value of 127.999. Since 128 is a power of 2, why does this happen?

7 Rewrite the IFDEMO program to find odd or even numbers using MOD instead of DIV.

CHAPTER 6

Procedures and Functions

Pascal Predefined Functions
Writing Your Own Functions
Calling of Procedures
Procedure Calls
Dummy Parameters and Calling Parameters
Return of Values From Procedures
Reference Parameters Versus Value Parameters
Arrays as Value and Reference Parameters
Recursive Programming
Recursion Versus Iteration
Exiting From Procedures and Functions
Forward Declaration of Procedures and Functions
Passing Procedures as Arguments
Problems
References

PASCAL PREDEFINED FUNCTIONS

We have already seen and used a few of the predefined functions of Pascal. Now we will specify them rigorously. A *function* is a subprogram that returns a single value of some type. The type may be any of the standard types: REAL, INTEGER, BOOLEAN, or CHAR. Functions written by users may return single values of some user-defined type as well.

Functions Returning a Real Value

ABS(X)	absolute value of X
ARCTAN(X)	arctangent of angle X (radians)
COS(X)	cosine of angle X (radians)
EXP(X)	exp(X)
LN(X)	natural log of X
LOG(X)	common log of X (not in all versions)
SIN(X)	sine of angle X (radians)
SQR(X)	X squared
SQRT(X)	square root of X

Run-time errors will occur for LN or LOG of negative or zero values, the SQRT of negative numbers, and the SQR or EXP of numbers that become too large to be represented.

Functions Returning an Integer Value

ORD(C)	integer value of character C
ROUND(X)	rounds the real number X and converts to integer
TRUNC(X)	truncates the real number X and converts to integer
ABS(I)	absolute value of integer I
SQR(I)	square of integer I
PRED(I)	integer I minus 1
SUCC(I)	integer I plus 1

Functions Returning a Character Value

CHR(X)	returns the character equal to the integer value X
PRED(C)	returns the previous alphabetic character
SUCC(C)	returns the next alphabetic character

Run-time errors will occur if the value X cannot be converted to a legal character code or if the character PREC(C) or SUCC(C) does not exist.

Functions Returning a Boolean Value

ODD(I)	returns TRUE if I is odd
EOF (file)	returns TRUE if scanning has reached an end of file
EOLN(file)	returns TRUE if next character is an end of line marker from file or terminal

WRITING YOUR OWN FUNCTIONS

In addition it is possible for you to write functions that return values for any given set of parameters. Let us consider the program we wrote to calculate factorials in Chapter 5. This might be more useful as a function, since a number of probability equations utilize several factorials.

To declare a function in a Pascal program, the first line must contain the function name, the names and types of all variables that it will use, and the type of value it will return. For example,

```
FUNCTION FACT(N:INTEGER):REAL;
```

declares that the function FACT will require one integer argument and will return a real value.

Since Pascal requires that all variables and other terms be declared in advance of their use, the function FACT must be written to stand ahead of the main program, or any other routine that calls it. In fact, if there are several functions or procedures, they all must precede the main program which, thus, comes last. This is simply a further element of structure in Pascal which requires you to define all of your routines before you use them.

Within the function there can be any desired number of *local* variables and constants that may have the same names or different names from those declared in the main program. Even if they have the same names, if they are declared again by a VAR declaration within the function, they are separate storage locations and have no relation to those in the main program. We say that the *scope* of these variables is local. At the end of the procedure's compilation, the compiler "forgets" these names and they could be reused for some other purpose. Naturally, using the same variable name in two contexts is poor programming practice and is to be discouraged. If, on the other hand, a function needs to make use of a variable that is part of the main program, this variable can be referred to and used within the function, as long as the variable name is not redeclared in the function. Such variables are then termed *global variables* and must be treated with care if their values are changed by the function. The proposed standard requires that the FOR-DO loop variable be *local* to the procedure or function in which it occurs, so that the loop will not inadvertently change the value of a global variable. This is particularly important since the value of the loop variable is undefined after successful execution of the loop. Use of such local variables is illustrated in the NMR programs that follow.

The most important variables that the function deals with are those within the parentheses of the function call itself. These are called the *calling parameters* or *arguments* of the function and can vary from one call to another, as shown in the example below. In addition, a function returns a single value at the place of its call.

```
P:= 2*FACT(N);   (*P IS 2 TIMES N-FACTORIAL*)
```

In the statement above, N is the calling parameter of the function FACT, and the real variable P is set equal to two times the factorial of N. The multiplication occurs after the evaluation of the function FACT.

Within the function, the final value of the function is specified by setting the function name equal to a number or expression, somewhere within the function:

```
FACT:=F1;   (*FINAL VALUE OF FUNCTION*)
```

Functions may only return values having simple types, such as CHAR, BOOLEAN, INTEGER, or REAL, as well as subrange types (Chapter 10) or pointer types (Chapter 13).

In the example program below, the function is named FACT and calculates the factorial value of the number N which is the calling parameter. It returns the factorial as a real number. In combinatorial arithmetic, the number of possible combinations of n things taken r at a time is given by:

$${}_nP_r = \frac{n!}{(n-r)!}$$

This equation is evaluated in the program below for values of n and r entered at the terminal. If $r >= n$, no evaluation takes place. Note that the expression (N−R) is enclosed within parentheses and is evaluated as the calling parameter for the second call of FACT. The value that FACT returns is determined by the statement

FACT:= expression;

at the end of the function.

```
PROGRAM COMBO;
(*THIS PROGRAM CALCULATES THE NUMBER
OF COMBINATIONS OF 'N' THINGS TAKEN 'R' AT A TIME
UNTIL  EITHER N OR R ENTERED IS A ZERO*)

VAR N,R:     INTEGER; (*MAIN PROGRAM VARIABLES DECLARED FIRST*)
    P:REAL;

(*****************************************************)

(*FUNCTION TO CALCULATE THE FACTORIAL OF ANY NUMBER N*)

FUNCTION FACT(N:INTEGER):REAL;
(*TAKES INTEGER ARGUMENT AND RETURNS REAL RESULT*)
VAR F1: REAL;    (*LOCAL VARIABLE*)

BEGIN
    F1:=1;       (*START WITH FACTORIAL OF 1*)

(*CALCULATE FACTORIAL UNTIL N IS REDUCED TO 0*)
    WHILE N > 0 DO
    BEGIN
        F1:=F1*N;
        N:=N-1;
    END;
    FACT:=F1;        (*RETURN FUNCTION VALUE*)
END;(*OF FACT*)

(*****************************************************)

(*MAIN PROGRAM BEGINS HERE*)
BEGIN
REPEAT
    WRITE('ENTER N<SPACE>,R<RETURN>: ');
    READLN(N,R);

(*CALCULATE FACTORIAL P=N!/(N-R)! *)

IF (N >= R) AND (R>0) THEN       (*BE SURE NO DIV BY 0 *)
    BEGIN
    P:=FACT(N)/FACT(N-R);
    WRITELN('P = ',P);
    END(*THEN*)
ELSE
    WRITELN( 'N MUST BE >- R!');
UNTIL (N=0) OR (R=0);    (*QUIT IF N=0 OR R=0*)
END.
```

CALLING OF PROCEDURES

Procedures are subprograms called by the main program and by other procedures that operate on arguments of the procedure and on other global variables. They return values to these arguments and variables and perform various calculations and input and output functions that may be needed at several points in a program. As an example of the use of a procedure, we consider the case of the calculation of nuclear magnetic resonance spectra (1), which involves getting the number of spins (N), the chemical shifts V[i], and the coupling constants J[i,j], and combining them to form a series of Hamiltonian matrices which are diagonalized to form a series of energy levels. The frequencies of the lines in the nuclear magnetic resonance spectrum are determined by the distance between energy levels and the intensities by relationships between chemical shifts and coupling constants. If we were to write an outline for such a program, we might write:

```
GETN;      (*GET NUMBER OF SPINS FROM TTY*)
GETVS;     (*GET N CHEMICAL SHIFTS*)
GETJS;     (*GET COUPLING CONSTANTS*)
CALCH;     (*CALCULATE HAMILTONIAN MATRIX*)
DIAGN;     (*DIAGONALIZE MATRIX*)
ENERGIES;  (*CALCULATE ENERGY LEVELS*)
   etc.


PRINTRESULTS;  (*PRINT OUT RESULTS*)
```

This program outline actually represents a list of *calls* to procedures by their names and thus is actually a structure for the program to be written. Such an outline represents a good example of top-down or structured programming, where the subroutines are all specified from the beginning of the design by their use in the planned program.

Here the names describe procedures to be written by the programmer as he proceeds with the task of defining the writing of the nuclear magnetic resonance program. Each of the procedures is then written separately to conform to this outline.

PROCEDURE CALLS

A procedure *call* is simply the name of the procedure itself, followed by the arguments in parentheses. A call to GETVS might be followed by the number of spins:

```
GETVS(SPINS);
```

Where the actual procedure GETVS begins, the name must be preceded by the word PROCEDURE and followed by the variables in parentheses by name and type:

```
PROCEDURE GETVS(N:INTEGER);
```

The procedure itself does not define a type of result, since no specific value is returned to the calling program from the procedure. Instead, the procedure returns values to some of the calling parameters, to some global variable, or just does input or output without changing any program variables at all.

Note that the name of the variable within the procedure need not be the same as in the calling program. In fact, it is generally useful if it does not have the same name to emphasize that the values passed to the procedure can be different variables at different times.

DUMMY PARAMETERS AND CALLING PARAMETERS

The data that we pass to the procedure from the calling program are referred to as *calling parameters* since they are the ones that the procedure uses in this call, and could vary from one call to the next within the same program. The variable names that the procedure itself uses, which correspond to the calling parameters, are called *dummy parameters* and are merely place holders, telling the compiler that this many variables will be passed from the calling program. They have no independent existence of their own and, as such, need not be declared as variables within the procedure. Thus we call the procedure GETVS with the *calling* parameter SPINS:

```
GETVS(SPINS);
```

in the main program, but we use a *dummy* parameter having the name N within the procedure itself:

```
PROCEDURE GETVS(N:INTEGER);
```

Pascal compilers always check to see that the calling parameters are of the same type as the dummy parameters, and they issue an error message if they do not agree.

RETURN OF VALUES FROM PROCEDURES

Within the procedure some calculations presumably use the values of the calling parameters. These calculations are useless, however, unless their results can be communicated back to the main program. There are two ways to communicate values back to the calling program: through the use of *global variables* and by *reference parameters.*

Global variables are simply names that are known to both the main program and the procedure by declaration in the first VAR declaration of the program. They are neither passed as calling parameters nor declared as local variables, but they are used by simply referring to them by the same name in both the main program and the procedure. Such global variables must not be redeclared by a VAR declaration in the procedure, or they will refer to separate values and not the same global values.

Thus rather than using the calling parameter approach, we could have a main program where we declare SPINS and then use it in the routine GETVS directly, rather than passing any arguments to GETVS. This approach is illustrated in the short example program NMR in which one procedure gets the value of SPINS and the next one allows entry of SPINS values of V(I). The main program then consists of only two short statements: a call to GETN and a call to GETVS. (The remainder of this obviously complex program is omitted for brevity.) Note the use of the local variable in the FOR loop of GETVS.

REFERENCE PARAMETERS VERSUS VALUE PARAMETERS

Pascal allows you to decide whether or not you want to allow the calling parameters to be changed by the procedure by the simple feature of preceding

```
PROGRAM NMR;
VAR SPINS:INTEGER;
    V: ARRAY[1..6] OF REAL;

(**********************************************)
(* PROCEDURE  FOR GETTING THE VALUE OF SPINS*)
PROCEDURE GETN;
BEGIN
    WRITE('NUMBER OF SPINS = ');
    READLN(SPINS);   (*GET NUMBER OF SPINS*)
END;(*GETN*)
(**********************************************)
(*PROCEDURE TO GET CHEMICAL SHIFTS*)

PROCEDURE GETVS;
VAR I: INTEGER; (*LOCAL VARIABLE*)
(*THE ARRAY V[I] IS DECLARED IN  THE MAIN PROGRAM*)
(*IT IS A GLOBAL  VARIABLE*)

BEGIN    (*GETVS*)
    FOR I:=1 TO SPINS DO     (*I MUST BE LOCAL ! *)

    BEGIN
        WRITE('V(', I:2, ')= ');
        READLN(V[I]);
    END; (*FOR*)
END;     (*GETVS*)
(**********************************************)

(*MAIN PROGRAM STARTS HERE*)

BEGIN    (*NMR*)

    GETN;    (*GET NUMBER OF SPINS*)
    GETVS;   (*GET THAT NUMBER OF SHIFTS*)
(*
    :
    :    *)
END.
```

those that the procedure may change by the VAR declaration within the procedure statement:

```
PROCEDURE GETN(VAR N:INTEGER):   (*N CAN BE CHANGED*)
```

All variables in a list of the same type can be changed if preceded by a VAR declaration:

```
PROCEDURE SNARCH(VAR A,X,K:REAL):
(*A,X,AND K CAN BE CHANGED*)
```

Those of a given type that are not preceded by VAR are copied, and only their values are passed to the procedure. The procedure can then change the values of those copied variables as many times as desired without causing any change in the original variables in the calling program:

```
PROCEDURE GETVS(N:INTEGER); (*N CANNOT BE CHANGED*)
```

This extremely powerful feature allows you to use important parameters as arguments to procedures without worrying that the procedure will inadvertently change them. Further it allows you to specify which ones you *expect* to see changed upon return from the procedure.

The terms for these two different types of parameters in procedures are *reference* parameters and *value* parameters. Value parameters cannot be changed in the calling program since only a copy of their *value* is passed to the procedure. Reference parameters can be changed in the main program by the procedure because the actual address where they are stored is passed to the procedure, and they can be *referred* to directly.

Pascal also allows any calling parameter in a procedure (or function) to be an expression:

```
SNARCH(X+Y, X, Z*Y);
```

although this sometimes reduces readability.

In the second version of the program NMR below, we see that the variable N becomes a reference parameter in the procedure GETN, since we *do* wish to return N to the main program. However, once we have determined a value for N, we do not wish it to change, and N is thus a value parameter in the routine GETVS and cannot change the main variable SPINS.

ARRAYS AS VALUE AND REFERENCE PARAMETERS

Since value parameters have only their values copied for use in a procedure, an array used as a value parameter will often require that the values of the entire array be copied into new memory for use in the procedure. This wastes both copying time and additional memory and is usually avoided. Arrays should in general be preceded by the VAR declaration in procedure definitions, so that they are always used as reference parameters. Single elements of arrays may be passed as either value or reference parameters, but elements of PACKED arrays (Chapter 10) cannot be passed as reference (VAR) parameters.

```
PROGRAM NMR2;
VAR SPINS:INTEGER;
     V: ARRAY[1..6] OF REAL;

(**************************************************************)
(* PROCEDURE  FOR GETTING THE VALUE OF SPINS*)
PROCEDURE GETN(VAR N: INTEGER); (*HERE N CAN BE CHANGED*)
BEGIN
     WRITE('NUMBER OF SPINS = ');
     READLN(N);   (*GET NUMBER OF SPINS*)
END;(*GETN*)
(******************************************** ******)
(*PROCEDURE TO GET CHEMICAL SHIFTS*)

PROCEDURE GETVS(N:INTEGER);
(*THIS PROCEDURE CANNOT CHANGE N IN THE MAIN PROGRAM*)
VAR I: INTEGER; (*LOCAL VARIABLE*)

(*THE ARRAY V[I] IS DECLARED IN
THE MAIN PROGRAM*)

BEGIN   (*GETVS*)
     FOR I:=1 TO N DO
     BEGIN
          WRITE('V(', I:2, ')= ');
          READLN(V[I]);
     END; (*FOR*)
END;     (*GETVS*)
(**************************************************************)

(*MAIN PROGRAM STARTS HERE*)

BEGIN   (*NMR*)

     GETN(SPINS);        (*GET NUMBER OF SPINS*)
     GETVS(SPINS);       (*GET THAT NUMBER OF SHIFTS*)
(*
     :
     :
     :   *)
END.
```

RECURSIVE PROGRAMMING

One of the most powerful features of Pascal is its *recursive* nature. Many of the definitions of the Pascal statements are themselves recursive in nature, meaning that they keep calling themselves until the expression becomes simple enough to evaluate. Recursion is an important technique in programming, and many modern computer languages support it to some extent. While the concept of a function or procedure calling itself seems like an endless loop with no exit, in reality recursion is a powerful technique to use whenever any programming task can be simplified by expressing it in terms of another problem involving fewer steps of the same type. An elegant and delightful description of recursion in life, art, music, and logic is given by Hofstadter in his major work, "Goedel, Escher, Bach: An Eternal Golden Braid" (2).

One of the simplest and most easily understood examples of recursion is the factorial problem we have already examined. The factorial of any number is simply that number n multiplied by $(n - 1) \times (n - 2) \times \ldots$ until $(n - x)$ becomes zero, where we then define 0! as 1. Putting this definition another way, $n!$ is equivalent to n times $(n - 1)!$, and we have thus defined a problem in terms of a simpler version of itself.

Use of functions or procedures that call themselves means that eventually the procedure must stop calling itself and begin exiting from all those calls, or we will indeed have an endless loop. In the case of the factorial, this should happen when the number operated upon becomes zero, since while we can define 0!, we cannot define $(-1)!$. Thus we have all the criteria we need for a simple recursive solution to the problem in which the function FACTR(N) calculates N*FACTR(N−1) until N becomes zero.

RECURSION VERSUS ITERATION

We have already seen that the factorial problem can be solved iteratively: by simply iterating until N becomes zero. The recursive solution is a bit simpler, however, and if well commented can be an invaluable programming tool.

```
PROGRAM FACTR2;
(*CALCULATES THE FACTORIAL
OF ANY ENTERED NUMBER >= 0*)

VAR
     NUM:INTEGER;
     X:REAL;
(**************************************************)
(*FUNCTION TO CALCULATE FACTORIAL RECURSIVELY*)

FUNCTION FACTR(N:INTEGER):REAL;
BEGIN
     IF N>0 THEN
          FACTR:=N*FACTR(N-1) (*NOTE FUNCTION CALLS ITSELF*)
     ELSE
          FACTR:=1; (*CAUSES EXIT INSTEAD*)
END;
(**************************************************)
(*MAIN PROGRAM BEGINS HERE*)
BEGIN
REPEAT
     WRITE('N= ');
     READLN(NUM);
     IF NUM >=0 THEN
          BEGIN
          X:=FACTR(NUM);   (*CALCULATE FACTORIAL OF N*)
          WRITELN(NUM,' FACTORIAL = ',X);
          END;
UNTIL NUM<=0;

END.
```

Iteration should be chosen whenever it seems simpler to you, and recursion whenever it seems that it might be possible to express a solution in terms of a simpler version of the same problem.

Be cautioned, however, that recursion may take up large amounts of temporary storage if a function must be called many times, since every new call to the function will require some additional memory for the saving of the various machine variables that exist during the current call. While this saving process is "invisible" to the Pascal programmer, the computer may run out of memory if a very large number of calls are made, and an error message such as STACK OVERFLOW may occur.

EXITING FROM PROCEDURES AND FUNCTIONS

Standard Pascal does not allow one to exit from the middle of a procedure or function in cases where it is determined that no further calculation is needed, but extensions to Pascal have been implemented in many compilers to permit this feature. For the most part these extensions are used to avoid the inherent structure of Pascal and are a programming convenience for those used to other languages.

In UCSD Pascal and DEC-10 Pascal, the EXIT(procedurename) command is available, which allows exit from a procedure or function at any point in the subprogram instead of requiring that exit be through the last END statement of the subprogram. This command can be used whenever an intermediate step of a calculation determines that the remainder of the calculation cannot be performed because of invalid data such as a zero denominator. However, the EXIT statement is not recommended since it makes it harder to follow the flow of the program compared to a few well-constructed IF statements. An example of the EXIT statement is given below in the hypothetical program having procedures A and B:

```
PROCEDURE A;
BEGIN
     B;   (*CALL PROCEDURE B*)
     :
     :
END;

PROCEDURE B;
BEGIN
     IF N=0 THEN EXIT(B) (*EXIT FROM THIS PROC*)
or alternatively
     IF N=0 THEN EXIT(A) (*OR EXIT FROM CALLING PROC*)
     :
     :
END;
```

```
(*MAIN PROGRAM*)
BEGIN
    A;  (*CALL PROC A WHICH IN TURN CALLS B*)
    :
    :
END.
```

In the above example the main program calls procedure A which in turn calls procedure B. Procedure B tests the variable N and if it is zero, it exits immediately from procedure B or alternatively could even exit from procedure A. EXIT is not defined in the proposed standard and is thus implementation dependent.

FORWARD DECLARATION OF PROCEDURES AND FUNCTIONS

When a program contains several procedures or functions which may call each other, there is sometimes a problem of which one should come first so that the compiler will know that a procedure exists when a call to it is found. Since the compiler scans the Pascal statements only once, it must know about the existence of any procedure or function when it encounters a reference to it. The problem of several procedures or functions which may call each other is overcome by the FORWARD declaration, which allows the programmer to declare that a procedure or function exists but has not yet been encountered.

```
PROCEDURE CHISQ(Y,YFIT:REAL);
FORWARD;    (*MEANS THE ACTUAL PROCEDURE OCCURS LATER*)
```

Then the actual procedure can be listed anywhere after this, but still before the main program. Where the procedure is actually entered in the program, the variable declarations in parentheses are *not* repeated:

```
PROCEDURE CHISQ; (*NO VARIABLE DECLARATIONS*)
```

Sometimes the variables are repeated inside a comment to make reading the program easier:

```
PROCEDURE CHISQ; (*Y,YFIT:REAL*)
```

PASSING PROCEDURES AS ARGUMENTS

Some versions of Pascal allow you to pass the names of procedures as arguments to other procedures. This is not allowed in UCSD Pascal, Aspect Pascal, or DEC-10 Pascal. In typical examples, where allowed, this features lets you declare a general type of function as one of the arguments of a procedure and then call the procedure with a specific function name and arguments embedded in the call:

```
PROCEDURE BISECT(FUNCTION F(X:REAL):REAL;
   A,B:REAL: VAR RESULT:REAL):
```

Note that there may be little difference between putting the actual function in the call and putting the value returned by the function in the call.

PROBLEMS

1 Write a function to return the value of y in the equation.

$$y = mx + b$$

for arguments of m, x, and b.

2 Write a procedure to find the largest element in a two-dimensional array and divide all elements by it.

3 Write a procedure to interchange the rows and columns of a square matrix.

4 The best straight line $y = mx + b$ having slope m and intercept b is given for a sequence of points x_i, y_i by calculating

$$\Sigma x_i, \quad \Sigma x_i^2, \quad \Sigma y_i, \text{ and } \quad \Sigma x_i y_i$$

for all n points. Then the slope m is given by linear least squares by

$$m = \frac{n \Sigma x_i y_i - \Sigma x_i \Sigma y_i}{n \Sigma x_i^2 - (\Sigma x_i)^2}$$

and the intercept by

$$b = \frac{\Sigma x_i \Sigma y_i - \Sigma x_i \Sigma x_i y_i}{n \Sigma x_i^2 - (\Sigma x_i)^2}$$

Write a program to read in a sequence of point pairs and then call a procedure LINLSQ to find m and b.

5 The largest common factor of two integers a and b is the largest integer that is a divisor of a and b. The function LCF (a, b) can be defined as

$$\begin{aligned} \text{LCF}(a, b) &= a, \qquad \text{if } b = 0 \\ &= \text{LCF}\ (b, a \text{ MOD } b), \qquad \text{if } b = 0. \end{aligned}$$

Write a recursive function LCF to evaluate the largest common factor.

REFERENCES

1 Edwin D. Becker, "High Resolution NMR; Theory and Chemical Applications," 2nd edition, Academic Press, New York, 1980.

2 Douglas Hofstadter, "Goedel, Escher, Bach: An Eternal Golden Braid," Basic Books, New York, 1979.

CHAPTER 7

Input and Output in Pascal

We have already made substantial use of the READ, READLN, WRITE, and WRITELN statements informally in previous chapters. We will define them rigorously in this chapter in terms of the basic input-output structure of Pascal.

CHARACTER INPUT AND OUTPUT

Pascal originally permitted only the reading and writing of characters, one at a time, through a file *buffer variable* (window pointer). All features for the input and output of integers and real numbers from text files were added later.

The standard file INPUT (or any other file that Pascal might define) was considered to have a window pointer INPUT↑ which contained the next character to be read. The first character of the file INPUT is automatically loaded into the file window when the program is started, and further characters are loaded with the function GET:

```
GET(INPUT);
```

which loads the next character into the file window. The function which modern Pascal performs to read a single character with the statement

```
READ(C);   (*C IS A CHAR VARIABLE*)
```

can also be written as:

```
C:=INPUT↑;   (*GET CHAR FROM WINDOW*)
GET(INPUT);   (*AND READ THE NEXT ONE INTO WINDOW*)
```

Similarly, the PUT statement moves one character from the file window to the output file:

```
WRITE(C);
```

is equivalent to:

```
OUTPUT↑:=C;
PUT(OUTPUT);
```

While we will not use the GET and PUT statements in character files, they will become important later in record files, and it is important to recognize the "look-ahead" capability that this file window gives you. Because of this window, we can always find out the characteristics of the *next* character of the input file.

CARRIAGE RETURNS AND SPACES

Unlike most other languages, Pascal ignores end-of-line markers or returns in strings of input data and treats them as spaces. For brevity in this discussion we will refer to this line-terminating marker character as a return, even though the actual "return" character may not be used in all systems. Returns may be inserted between numbers entered in a READ statement, even if they are not required by a READLN statement. However, if you want to make sure that a carriage return is entered in the data stream, you can require it with a READLN statement, which continues reading characters until a return is found, even if it must bypass other numbers or characters to get there. This, then, provides a way of skipping over irrelevant data in a file when desired.

Thus the statement

```
READ(A,B,C);
```

will read the first three legal numbers entered into variables A, B, and C, regardless of whether A and B or B and C are separated by spaces, tabs, or returns, while the statement

```
READLN(A,B,C):
```

will read the first three legal numbers entered into variables A, B, and C and continue looking for characters until a return is found, even if C is terminated with a space or a tab.

THE EOLN FUNCTION

Since returns are actually ignored in much of Pascal, the function EOLN can be used to see whether a return has been entered. The function

EOLN(INPUT)

will return the Boolean value TRUE if the next character is a return and FALSE at all other times. Further, the EOLN function can be used on any text file or input from the terminal in the same way:

EOLN(fileidentifier)

If no argument is supplied to EOLN, the INPUT file is assumed. The following statements read characters into an array until a return is typed, where A is an array of CHAR:

```
I:=1;  (*SET ARRAY INDEX TO START*)
WHILE NOT EOLN DO
   BEGIN
   READ(A[I]);
   I:=I+1;     (*NEXT ELEMENT*)
   END;
```

The EOLN function can be used to allow entry of new values of parameters or keep old or default values, depending on whether a return or a number is entered. The following program fragment prints out the current value of WIDTH, a space, and allows entry of a new value or a return. The old value of WIDTH is kept if only a return is entered.

```
WRITE('WIDTH = ',WIDTH,' ');      (*PRINT PROMPT MESSAGE*)

(*NOW GET EITHER A NEW NUMBER OR A RETURN*)
IF EOLN
     THEN READLN     (*SKIP OVER RETURN IF NO DATA*)
ELSE
     READLN(WIDTH); (*IF NOT EOLN THEN READ IN NEW VALUE*)
```

FORMATTING OUTPUT TO THE TERMINAL OR PRINTER

Up to this point we have been using the WRITE and WRITELN statements and listed all the variables and messages we wish to print out inside parentheses without regard to format. Data in Pascal are printed out in a default format for each data type, unless overridden by a specific format specification. The default format varies with the system and should be checked for each installation.

Integer Default Format

In UCSD Pascal, integers are printed using exactly the number of spaces needed for that size of integer. A sign is included only if the integer is negative.

In DEC-10 Pascal, integers are printed in a 12 column format unless a specific format is specified.

Real Number Default Format

In UCSD and DEC-10 Pascal, real numbers are printed out in a 12-column expoential format unless a format is specified. Numbers between 1 and 10 will be printed in a decimal format with nine decimal places.

THE PAGE STATEMENT

New pages can be started in the output of any program by simply including the statement

```
PAGE;
```

This generates a formfeed character which causes a page advance in most printers, although it may not have any effect in many slower terminals. Other character files can also have formfeed characters inserted with

```
PAGE(file);     or     PAGE(OUTPUT);
```

(UCSD Pascal requires the file identifier in PAGE statements.)

FORMATTING REAL AND INTEGER NUMBERS FOR OUTPUT

To make an integer occupy a known number of columns, simply follow it in the WRITE statement with a colon and an integer or constant name equal to the number of columns you wish the number to occupy:

```
WRITELN(AINT:8, BINT:10);
```

The above statement will fill with spaces from the left, assuring that AINT will occupy eight columns and BINT will occupy the next ten on the same line.

If the specified format is too small for the integer, Pascal will override it and print out the minimum number of positions that is adequate. In calculating the space needed, note that Pascal will include a space for a sign if the number is positive and print a minus sign if it is negative.

FORMATTING REAL NUMBERS

There are two parameters describing the format of real numbers: the total number of columns to use and the number of places after the decimal point. If no format is specified, Pascal will print out the number in a default exponential format, where

5.2E7 means 5.2×10^7

If you wish to specify a format, you can specify the total fieldwidth only:

```
CONST
   FIELDWIDTH=10;
VAR
   A:REAL;
BEGIN
   WRITELN(A:FIELDWIDTH);
```

or

```
   WRITELN(A:10);
END;
```

in which case the value will be printed out in scientific notation (E format).

Alternatively, you can specify both the fieldwidth and the number of decimal places:

```
CONST
   FWIDTH=16;     (*TOTAL FIELD WIDTH*)
   DECS=4;        (*PLACES TO RIGHT OF DECIMAL*)
VAR
   A:REAL;
BEGIN
   WRITELN(A:FWIDTH:DECS);
END;
```

which will cause the decimal format to be used where FWIDTH defines the total number of places and DECS defines the number to follow the decimal point. Again, if the fieldwidth is too large it will be filled with spaces from the left, and if it is too small, Pascal will expand to a convenient format, usually the E format. (DEC-10 Pascal does not currently expand to a correct format.)

Note that in the above example we have put the fieldwidth and decimal place count in the constant declaration section so that it will be constant throughout the program. This saves us having to remember what the numbers are when we write the rest of the program, and enables us to change them easily if we want to change the entire format at once by simply changing the definitions of the two constants. This leads to a more organized or structured approach to programming and permits more efficient debugging and program maintenance. These format values may be represented symbolically as shown, and may be constants, variables, or expressions.

The following program illustrates the use of various format specifications. The DEC-10 output follows the program. Note that current DEC-10 Pascal does not automatically expand inadequate formats, but only prints out a pair of asterisks. Further it does not "round up" the output values as required by the proposed standard.

```
PROGRAM FORMAT;
(*THIS PROGRAM ILLUSTRATES THE USE
OF VARIOUS FORMAT SPECS*)

CONST A=99.63;   B=768;
FWIDTH=10;   DEC=3;   (*FORMAT CONSTANTS*)
BEGIN
(*FIRST WRITE A STRING TO SHOW COLUMN COUNT*)
WRITELN('12345678901234567890');
    WRITELN(A); (*DEFAULT REAL FORMAT*)
    WRITELN(B); (*DEFAULT INTEGER FORMAT*)
    WRITELN(A:FWIDTH);   (*SINGLE SPEC FOR REAL*)
    WRITELN(B:FWIDTH);   (*SINGLE SPEC FOR INTEGER*)
    WRITELN(A:FWIDTH:DEC);   (*WIDTH-DEC FORMAT FOR REAL*)

(*NOW TRY TO PRINT IN TOO SMALL A SPACE*)
    WRITELN(A:3:2); (*NEEDS AT LEAST 4 POSITIONS*)
    WRITELN(B:2);    (*NEEDS AT LEAST 3 FOR INTEGER*)
WRITELN('12345678901234567890');
END.
```

DEC-10 Output:

```
1234567890567890
 9.962999939E+01
           768
 9.962E+01
        768
    99.629
**
**
123456789034567890
```

UCSD Pascal Output

```
12345678901234567890
 9.96300E1
768
 9.96300E1
       768
    99.630
 99.63
768
12345678901234567890
```

FORMAT OF INPUT DATA IN PASCAL

Standard Pascal, which includes UCSD Pascal and DEC-10 Pascal, requires that if an integer is expected, an integer must be entered, and if a real number is to be read in, a real number must be entered. This means that integers must contain no decimal points or fractional parts and real numbers *must* have digits on both sides of the decimal point or on both sides of the E. Thus

355	is an integer
27.6	is real
92E3	is real
34.	is illegal for either type

This distinction is important in that the entry of a number of the wrong type may either abort the program immediately or cause lengthy error messages. Aspect Pascal is more forgiving and automatically converts the data to the proper type. A "forgiving" REAL input routine is shown in Chapter 16. The proposed standard and many compilers terminate input at the first character illegal for that type. Thus a decimal point will terminate an integer READ and the decimal point will be left to be read by the next READ. This, of course, still could be disastrous if an integer is expected.

THE TERMINAL AS AN INTERACTIVE FILE IN UCSD PASCAL

In UCSD Pascal the terminal is considered the primary input device and becomes the files INPUT and OUTPUT. Since the look-ahead used by static files already created is much simpler than in the case of terminal input and output, the terminal is considered to be a special file of type INTERACTIVE. This file has no look-ahead characters: the function EOLN becomes true only if the return was the immediate previous character typed. Further, there is no line buffer associated with the terminal's input, and characters are read and processed immediately as called for. Thus the program fragment

```
WRITE('ENTER ONE CHARACTER: ');
READ (C);
```

will cause continuation of execution as soon as a single character has been typed, even though it is not terminated by a return.

TERMINAL I/O IN DEC-10 PASCAL

In DEC-10 Pascal the terminal look-ahead character is called INPUT↑ and can be accessed as a character variable. When data is entered followed by a return, the READLN statement would normally wait to get the first character of the next line as the new look-ahead character for the INPUT↑ window. When the input device is the terminal, the Pascal operating system delays getting this character so that additional data need not be typed at once, and terminal I/O thus appears to be the same as I/O from other files.

One difference from UCSD Pascal, however, is that single characters are not passed to the Pascal program until the current line is terminated with a return. Therefore

```
READ(C);
```

will not actually read a value from text being entered until both the character and the return are typed. The return character is not skipped, however, and will be passed as a space when the next character is read. Note by comparison that

```
READLN(C);
```

will read the character and move on to the beginning of the next line, skipping over the return character.

PROBLEMS

1 Write a program to read in an array of characters, all entered on one line, and print them out in reverse order on the next line.

2 Write a program to read in words having any number of spaces or returns between them and print them out one space apart, five per line.

3 Pascal's triangle describes the coefficients of the expression $(a + b)^n$ and the intensity of nuclear magnetic resonance splitting patterns. Each entry is the sum of those diagonally above it:

```
        1
      1   1
    1   2   1
  1   3   3   1
1   4   6   4   1
```

Write a program to print out 10 lines of Pascal's triangle.

4 Write a program to print out the numbers from 1.1 to 9.9 diagonally across the page. In other words, each new value should be five positions further to the right on the next line.

CHAPTER 8

Documenting Pascal Programs

INTRODUCTION

It has been noted that an undocumented program is of no value whatsoever, and if a programmer dies, such programs are usually buried with him. In order to assure yourself that such debris are not buried with you, it is essential that any program you wish to outlive you be well described and commented. Then a new user, or the programmer himself a month later, will be able to relearn the features of the program quickly.

We introduce this topic here since most of the fundamentals of Pascal programming have been covered, and it is now that you will begin writing substantial programs, either as assignments or for your own use. It is essential that you form good documentation habits from the start.

TYPES OF DOCUMENTATION

There are two major types of documentation useful in programs in all languages: the instructions for use and the comments within the program itself. These differ substantially in intent, although not necessarily in quantity.

The comments within the program are of the "how it works" variety, allowing the programmer and other users to see how the functions the program actually performs are carried out. Such comments usually start with one big block of descriptive comments, followed by smaller comments after or between many lines of the code.

The instructions-for-use document is much more fundamental in approach, describing the use of the program to the relative novice who may barely know how to begin communication with the computer at all. Both types of documentation are most important for the finished program, and programming assignments should not be considered complete until both have been prepared.

COMMENTING PASCAL PROGRAMS

Most Pascal program examples in this text are fairly heavily commented compared to program examples in other books, since it is our belief that the learning of programming by example should include all types of examples needed to write good programs. However, even here we have made some comments shorter than you might otherwise, simply for reasons of space. Following are a few general rules that apply to comments in Pascal programs, and to a large degree to comments in other languages as well.

1 Each program should begin with a long block of text describing exactly what the program is for and how it goes about doing it. It should include a date or version number and the author's name.

2 Each major function or procedure should be set off from the surrounding program code with a line of stars, dashes, number signs, or other unique characters, enclosed, of course, in a comment.

3 Each procedure or function should have its own comment block describing how it works, what each dummy parameter and argument is, and noting any global variables that are used.

4 Within the procedure and the main program, comments should precede each logical section of the program, describing what that section does.

5 After any line of code doing something important, a short comment should be included to show that this is the place where the operation occurs. It is also important to explain *why* things are done.

THOROUGH COMMENTING OF THE FACTR2 PROGRAM

To illustrate how thorough comments should really be in a finished program which must stand by itself, let us consider the simple program FACTR2 given in Chapter 6, which calculates factorials of any entered integer, and exits if the integer is < 0.

Our comments include the title, author, and date at the top of the program, a block of comments pertaining to the entire program, and a block describing the action of the function FACTR. In addition we see that more comments are added to the main program to make clearer what is going on. Finally, we also note that this program has been somewhat altered to check more carefully for error conditions. Most of these changes come under the category of human engineering of the program to make it more foolproof.

HUMAN ENGINEERING OF PROGRAMS

Human engineering of programs means making them respond in a helpful manner to conditions that might otherwise produce puzzling results or run-time errors. It also means making them prompt the user for the needed data and having them issue nonfatal error messages for invalid data when possible. There is nothing more annoying than having just entered 20 or 30 numbers at the terminal and then having the program tell you that you made an error in entering the third item, or worse yet, having the program tell you nothing, but produce ridiculous answers instead.

Some fundamental tenets of a well-engineered program for human rather than machine use include:

1 Having the program print out a helpful prompt message for each datum required
2 Having the program print out error messages immediately upon encountering invalid data, and ask for a new value
3 Having all numerical output in a convenient format rather than the default exponential format
4 Allowing the user to change individual entries in a list of data after they have been entered, if this might improve the calculated result

In the FACTR2 program in Chapter 6, we see that there are some small inconsistencies which we ought to clean up if the final program is to be easy to use and understand. In particular we note that the program will allow entry of data that will produce factorials so large that they cannot be represented by values less than 10^{37}. We can easily protect the program against this by testing it, finding where overflow occurs, and preventing numbers larger than this from being sent to the FACTR routine. This might include the statement:

```
IF NUM < 33 THEN
```

Second, we note that we use entry of a number less than 0 as our condition to exit from the program, but that we do not indicate the reason for exit. We thus insert the additional IF condition that prints out an integer message that is too large or an illegal integer message at the end of the loop.

```
IF NUM>33 THEN
 WRITELN('VALUE TOO LARGE TO CALCULATE');
IF NUM<0 THEN
 WRITELN('EXITING, CANNOT TAKE FACTORIAL OF NEGATIVE NUMBER');
```

```
(*>>>>>*)         PROGRAM FACTR3;      (*<<<<<*)
(**************************************************

******** FACTORIAL CALCULATION PROGRAM ***********
           BY BLAISE PASCAL
             LAST REVISED 2/7/89
**************************************************
THIS PROGRAM ALLOWS THE TELETYPE ENTRY OF ANY INTEGER
AND CALCULATES ITS FACTORIAL. NEW NUMBERS CAN BE
ENTERED UNTIL  ONE LESS THAN ZERO IS ENTERED.
THIS CAUSES EXIT FROM THE PROGRAM.
    THE MAIN PROGRAM PROMPTS THE USER WITH "N="
AND ALLOWS ENTRY OF AN INTEGER, CALLS THE FUNCTION FACTR
TO CALCULATE THE FUNCTION RECURSIVELY AND PRINTS OUT THE
ANSWER UNLESS THE ENTERED NUMBER IS < 0.
IF IT IS, IMMEDIATE EXIT OCCURS.
(*************************************************

(****  VARIABLES DECLARED HERE              ***)
VAR
    NUM:INTEGER;
    X:REAL;
(**************************************************)
FUNCTION FACTR(N:INTEGER):REAL;

(* THIS FUNCTION CALCULATES THE FACTORIAL OF THE NUMBER
INTEGER N PASSED TO IT AS A VALUE PARAMETER
THE VALUE IS CALCULATED RECURSIVELY, MEANING THAT
THE FUNCTION CONTINUALLY CALLS ITSELF UNTIL THE NUMBER
N IN THE CALL IS ZERO. AT THIS POINT, INSTEAD OF A NEW CALL,
THE FUNCTION VALUE IS SET TO 1.0 AND EXIT OCCURS FROM
THIS FUNCTION AND SUBSEQUENTLY FROM ALL PREVIOUS CALLS.
WHILE RECURSIVE CALLS TAKE UP LARGE AMOUNTS OF
MEMORY SPACE, THE MAIN PROGRAM LIMITS THE ENTRIES
TO NUMBERS LESS THAN 33, SINCE FACTORIALS LARGER THAN
THIS CANNOT BE REPRESENTED IN THIS COMPUTER SYSTEM.    *)

BEGIN
    IF N>0 THEN (*CALL FUNCTION AGAIN UNTIL N=0 *)
        FACTR:=N*FACTR(N-1) (*NOTE FUNCTION CALLS ITSELF*)
    ELSE
        FACTR:=1; (*CAUSES EXIT INSTEAD*)
END; (*FACTR*)
(**************************************************)

(*MAIN PROGRAM BEGINS HERE*)
BEGIN
REPEAT
    WRITE('N= ');     (*PROMPT USER TO ENTER VALUE*)
    READLN(NUM);      (*READ IN INTEGER*)

(* ALLOW CALCULATION OF ALL FACTORIALS IN THE RANGE

        0 <= N <= 33

THOSE LESS THAN 0 ARE MEANINGLESS AND THOSE GREATER THAN 33
CANNOT BE REPRESENTED BY THIS COMPUTER'S REAL FORMAT *)
    IF (0<=NUM) AND (NUM<=33) THEN
        BEGIN
        X:=FACTR(NUM);   (*CALCULATE FACTORIAL OF N*)
        WRITELN(NUM:3,   ' FACTORIAL = ', X:10);
        END;(*IF*)
    (*GENERATE ERROR MESSAGES IF NUM<0 OR NUM>33*)
    IF NUM>33 THEN
        WRITELN('FACTORIAL TOO LARGE TO CALCULATE');
```

```
    IF NUM<0 THEN
        WRITELN('EXITING, CANNOT TAKE FACTORIAL OF NEGATIVE NUMBER');

UNTIL NUM<0;     (*QUIT CALCULATION IF NUM IS LESS THAN 0*)

END.
```

DOCUMENTATION OF THE FINISHED PROGRAM

Even though you now have a finished, working, well-commented program, it is useless unless you can provide some sort of documentation describing it. Documentation is the weakest link in programming projects at all levels from student to professional and often makes entire programs worthless, since no one knows exactly how to use them.

It is tempting to write a "self-documenting" program that prints out all instructions for its use, but such documenting is poor practice (a) because it presumes that the user knows how to start the program to get these instructions, (b) because they may be tediously long and boring to watch being typed out, and (c) because until the instructions are in hand, the prospective user has no idea whether the program will even do what he wants.

Instead, we suggest that you prepare a short document for even the most trivial of programs each time you write one, explaining what the program does and how to use it. A typical form of such a document might include the following:

1. Title of the program, author, and date of last revision.
2. A brief description of what the program does.
3. Necessary data and hardware for running the program.
4. How to start the program and how to tell that the program is indeed running.
5. Error conditions and what to do about them.
6. How to run more data.
7. Command conventions, if any.
8. How to abort a command or change data.
9. Detailed explanation of all commands.
10. Examples of use.
11. How to exit from the program.

In preparing a set of instructions such as this, keep in mind that you are writing for the novice or noncomputernik who will be put off by excessive use of jargon, as well as by inelegant grammar. You should do your best to write instructions that are as clear and concise as possible. Three general rules will help you write instructions as clearly as possible:

1 **Avoid Using the Impersonal.** Rather than writing

  "One must be sure not to enter a decimal point . . ."

  write

  "You must be sure that you don't enter a decimal point . . ."

2 **Use the Active Rather than the Passive Voice.** While this lends a certain nonscientific informality to the document, surveys have shown that it makes it much more readable. Thus rather than writing

  "When starting a return must be typed . . ."

  write instead

  "Type a return when you start the program . . ."

3 **Avoid Using "Input" and "Output" as Verbs.** This usage is particularly offensive to the noncomputer expert. Grating, inelegant words such as "inputted" or "outputting" are extremely "off-putting," and should be replaced with "entered" and "printed" (or "typed" or "displayed") instead.

Following is such a set of instructions for the FACTR3 program:

FACTR3
by Blaise Pascal
Last Revision 2/7/89

This program is designed to calculate the factorials of numbers between 0 and 33 and print them out on the terminal. The numbers must be entered at the terminal one at a time during the program.

The program requires a time-sharing computer or minicomputer system which runs a Pascal compiler. No stored data files or special hardware are required.

To start the program, establish communication with your computer system and type (for example)

```
EXE FACTR3      followed by a return
```

Pascal will compile and load the program, typing out compilation and loading messages. It will then begin execution by typing an asterisk, indicating that it is ready to begin. Type a return and the program will type out

N=

asking for a number between 0 and 33 to be entered. Enter an integer in this range followed by a return. The factorial of that number will be printed out in exponential form:

N=
4 FACTORIAL = 2.400E+01

where this means 2.4 times 10 to the first power. Exponential (E) notation is used because of the wide variation in the size of factorials that can be printed out. The program will then ask for a new number to be entered by typing "N=" again.

The program will type the error message

FACTORIAL TOO LARGE TO CALCULATE

if an integer greater than 33 is entered. A run-time error will occur if a number containing a fractional part is entered.

Exit from the program occurs when a value less than zero is entered. The message

EXITING, CANNOT TAKE FACTORIAL OF NEGATIVE NUMBER

is printed, and the program exits to the monitor.

Examples of Use on the Decsystem-10

(Underlined parts are typed by user.)

```
EXE FACTR3
PASCAL: FACTR3 [FACTR3    ]

  0 ERROR(S) DETECTED

LINK:    Loading
[LNKXCT FACTR3 Execution]

*
N=4
  4 FACTORIAL  = 2.400E+01
N= 33
 33 FACTORIAL  = 8.683E+36
N=34
FACTORIAL TOO LARGE TO CALCULATE
N=0
  0 FACTORIAL  = 1.000
N=-1
EXITING, CANNOT TAKE FACTORIAL OF NEGATIVE NUMBER

EXIT
```

CHAPTER 9

Files in Pascal

Text Files
Reading a Text File
The EOF and EOLN Functions
Other Types of Files
Passing File Variables to Procedures
Problems
References

The term *file* in Pascal refers to a series of data usually on some external storage device such as a disk or magnetic tape. Files are treated as if they were on a tape in that they can only be read sequentially and must be written from the beginning or after the last WRITE; they can never have data inserted in the middle.

TEXT FILES

In most versions of Pascal, the *text* file is the most common file type. It is simply a series of characters in a file which can be interpreted by reading them just as if they were entered from the terminal. The standard files INPUT and OUTPUT are both text files. It is, however, possible to have other files of this type, provided that they are declared at the beginning of the program.

In order to read or write data into such a file, it is necessary to define a *file variable* as having type TEXT, meaning that the file will be a text file:

```
VAR
   F1:TEXT;      (*TEXT FILE VARIABLE*)
```

The file variable is simply a variable name which we will use throughout the program and which usually is independent of the file's actual name on the external storage device. It may have any number of characters just as any other variable and, as usual, only the first 8 to 10 characters are significant.

Then, to specify the name that the file will have in the disk directory in UCSD or DEC-10 Pascal, the REWRITE statement is used, followed by the file variable and a file name in apostrophes:

```
REWRITE(filevariable, 'filename ');
```

or more specifically:

```
REWRITE(F1, 'FUMP   ');
```

The REWRITE statement defines a new file to be written onto the disk having the name FUMP and the file variable name F1. We use this file variable name just as we used the name OUTPUT in our READ statements to specify what device we are writing to. If instead of OUTPUT, the compiler encounters a known file variable name, the data will be written into that file instead.

Throughout this book we will use simple file variable names, such as F or F1. However, you should recognize that such variable names may have any number of characters and can start with any letter, just as other types of variable names can.

In DEC-10 Pascal the file name must be a string exactly nine characters long. If the filename is shorter, spaces must be added to the end. The first six characters are used to define the filename and the last three to define the extension, if any. A period may optionally be used between the filename and the extension, but there must be six characters or spaces before the extension.

In UCSD Pascal the number of characters in the filename string can be variable (up to 15), and the name is automatically filled with spaces (1). Then to actually write the data into the file, we use the WRITE statement with our file variable name instead of OUTPUT:

```
WRITTEN(F1, A, B);    (*WRITES VALUES A AND B TO FILE F1*)
```

This statement writes the variables A and B into the file F1 whose name we have defined above as FUMP. The data is written as a series of characters, just as it would be printed on the terminal.

READING A TEXT FILE

Once you have written a file onto disk, it is only useful if you can reread it into a Pascal program. In order to read the file, you must set the file's "pointer" back to the beginning of a file. Resetting the file's pointer is done with the RESET statement.

If you have just written a file with a series of WRITE commands to the file whose file variable is F1, you write

```
RESET(F1);
```

to put the file text pointer back to the beginning of the file. If you wish to begin reading from a file written by another program, you must define its filename in the same way as the REWRITE statement used before you can begin reading from it:

```
RESET(F1,'FUMP   ');
```

Then to read from the text file, you simply use the READ and READLN commands, as usual, with the device name equal to the file variable name instead of INPUT:

```
READ(F1, A, B);   (*READS VARIABLES
                   A AND B FROM FILE F1*)
```

It is not possible to read beyond the end of a file—spaces will be returned for all READs. Some time-sharing versions of Pascal (and the proposed standard) cause a run-time error after repeated attempts to read beyond the end of a file. DEC-10 Pascal allows eight attempts to read beyond the end of a file before the run-time error occurs. UCSD Pascal always returns characters READ beyond the end of a file as spaces, and reads real numbers and integers as zeros.

In the program FILETEST below, the values of real variables A and B and integer variable R are obtained from the keyboard, written into the file FUMP, and then reread and printed out for verification. Upon exit from the program, the file FUMP will contain the text characters equivalent to the printed out values of A, B, and R.

Note that the format of the data when written must be such that it can be reread. Thus the numbers must be separated from each other by spaces or returns. In this example we include spaces between A, B, and R so that, regardless of their size or sign, they can be read back properly. Small numbers, of course, will usually have leading spaces, and additional spaces are superfluous.

```
PROGRAM FILETEST;
(*THIS PROGRAM WRITES DATA INTO THE DISK FILE 'FUMP'
REREADS IT AND PRINTS IT OUT*)

VAR
     A,B:REAL;
     R:INTEGER;
     F1:TEXT;      (*TEXT FILE*)
(**********************************************)
PROCEDURE ABWRITE;
(*WRITES OUT CURRENT VALUES OF A, B, AND R*)
BEGIN
WRITELN('A=',A);
WRITELN('B=',B);
WRITELN('R=',R);
WRITELN;     (*SKIP A LINE*)
END;
(**********************************************)
```

```
BEGIN
WRITE('ENTER A,B (real), AND R(integer): ');
READ(A,B,R);

(*WRITE THEM OUT FOR VERIFICATION*)
ABWRITE;     (*PROCEDURE DUMPS THESE ONTO TERMINAL*)

(*NOW WRITE THEM INTO A TEXT FILE ON DISK*)
REWRITE(F1,'FUMP      ');     (*FIRST CREATE THE FILE*)
WRITELN(F1,A,' ',B,' ',R);

(*RESET THE FILE TO THE BEGINNING AND REREAD THE DATA*)
RESET(F1);
READ(F1,A,B,R);
ABWRITE;     (*TYPE THEM OUT AGAIN*)
END.
```

THE EOF AND EOLN FUNCTIONS

When you are reading data from a file of indeterminate length, it is useful to be able to find out whether there is more data in the file at any given time. The EOF (end of file) function provides this capability as a look-ahead feature. When the last character of a file is read, the function

EOF(filevariable);

becomes TRUE. This is comparable to your chess opponent calling out "check" when his *next* move may cause the end of game (checkmate). Thus you can read all the characters in a text file and get no spurious ones by checking the function EOF before each new READ. If you read beyond the end of file inadvertently, EOF will remain TRUE, and according to the proposed standard, a run-time error will occur. UCSD Pascal will return spaces for additional character READs beyond the end of file and zeros for integer and real READs. In DEC-10 Pascal only eight characters beyond the EOF can be read without causing a run-time error. All real and integer READs beyond EOF cause errors in DEC-10 Pascal, since no integers have been found after eight READs.

In a similar fashion, the EOLN function provides a look-ahead capability for files such that EOLN becomes TRUE when the last character before the return is read. The next character, the return itself, is returned as a space, and EOLN then becomes FALSE, since it is testing the *next* character—the one at the start of the next line. When EOF(file) is TRUE, the function EOLN(file) also becomes TRUE for some systems.

Thus to read a list of numbers from a text file into array X, you could write:

```
I:=1;     (*START AT BEGINNING OF ARRAY*)
WHILE NOT EOF(F1) DO      (*READ UNTIL EOF*)
     BEGIN
     READ(F1,X[I]);        (*READ A NUMBER*)
     I:=I+1;
     END;
N:=I-1; (*SAVE NUMBER READ*)
```

OTHER TYPES OF FILES

While the original Pascal report recognized only text files and record files (Chapter 11), the proposed standard allows files of any simple type which are not in themselves files. These are declared by using the FILE OF *type* statement:

```
VAR
F2: FILE OF REAL;
```

Further, the proposed standard permits reading and writing of these files using the READ and WRITE statements:

```
VAR
R:REAL;
F2:FILE OF REAL;
BEGIN
:
:
WRITE(F2, R);   (*WRITE REAL NUMBER INTO FILE F2*)
```

The procedures READLN and WRITELN are only allowed for text files, however. Many current compilers only allow such files to be read using the GET and PUT statements (Chapter 11).

PASSING FILE VARIABLES TO PROCEDURES

In Pascal the file variable itself may be passed as an argument to a procedure or function if the files are of the same type. Thus the terminal or INPUT file or some other text file may be specified to a procedure by passing a file variable to that procedure. In the definition, the file variable name and type must be specified:

```
PROCEDURE WRITEX(F:TEXT; X:REAL);
BEGIN
   WRITELN(F, 'X=', X);
END: (*WRITEX*)
```

Then the data can be written into the file F, whether it is the terminal or a defined file. An example of a complete program illustrating this feature is given in the program READR in Chapter 16.

PROBLEMS

1 Write a program to read in a text file and count the number of A's, C's, and L's in it. Print out the result.

2 Write a program to read in two files and combine them in a new file.

3 Write a program to read in a file of numbers and print them out one at a time. Then allow modification of each number by allowing the user to type S, D, or C for save, delete, or change before proceeding to the next number. Write a new file with these numbers modified.

REFERENCES

1 K. L. Bowles, "Beginner's Guide to the UCSD Pascal System," BYTE/McGraw-Hill, New York, 1980.

CHAPTER 10

More Complex Data Types

Thus far we have seen that data can be INTEGER, REAL, BOOLEAN, CHAR, or TEXT file. We see in this chapter that Pascal allows us to define any data type we like as a combination of these simple types and others.

THE TYPE DECLARATION

The TYPE declaration provides an economical and compact way to define data structures such as packed and normal arrays as well as the ranges of values that integer or character variables can take on. Such declarations must come after the CONST section and before any VAR declarations.

For example, the data type XARY can be defined as an array of 512 elements:

```
TYPE
    XARY=ARRAY[1..512] OF REAL;
```

Note the use of the equal sign without the colon. Then the variables of this new type can be defined as XARYs:

```
VAR
    X,Y,Z: XARY; (*ALL ARE 512 PT REAL ARRAYS*)
    :
    :
    PROCEDURE SLUNCH†(VAR A:XARY);
```

SUBRANGE TYPES

A more important use of the TYPE declaration is the subrange type, however. Most integers in a program can really take on only a limited range of meaningful values; for example, array subscripts are limited to the size of the array. Thus we often find in good Pascal programs that few variables are of the simple type INTEGER, since they can more correctly be declared as lying in a smaller range. Such variables are said to belong to a *subrange* type. The advantage to this type is that the compiler can then check to see that you have used the variables correctly and that during the running of the program, checks will be made for the validity of any such variable each time a new value is encountered. Thus Pascal helps you with both program construction and problems that might only be found through substantial amounts of debugging.

A subrange type is declared by the TYPE declaration followed by an equal sign and the bounds, much as an array is:

```
TYPE
     SUBSCRIPT=1..4096; (*CAN ONLY TAKE ON THESE VALUES*)
```

Then once the type has been declared, it can be used to describe variables:

```
VAR
     I,J:SUBSCRIPT;   (*I AND J CAN ONLY LIE BETWEEN 1 AND 4096*)
     X:ARRAY[SUBSCRIPT] OF REAL; (*4096-POINT REAL ARRAY*)
```

Subrange type declarations can also apply to characters, although they are less often used:

```
TYPE
     HIALPHA='Q'..'Z';
     LOWALPHA= 'A'..'P';
```

Subrange declarations can not be applied to real variables, however.

TYPE COMPATIBILITY

Once we have defined some variables having subrange types and others having a full range (INTEGER, CHAR) or another subrange, we might wonder how the compiler and run-time system decide if the types are compatible.

Variables have *identical* types if they are declared as having the same type:

```
TYPE
     SUB1=9..23; (*SUBRANGE TYPE*)
VAR
     A,B:SUB1;     (*A AND B IDENTICAL TYPE*)
```

```
TYPE
     SUB1=9..23;
     SUB2=SUB1;    (*EQUIVALENT TYPES*)
VAR
     A:SUB1;
     B:SUB2;       (*A IDENTICAL TO B*)
```

Types are defined as *compatible* by the compiler if one is a subrange of the other, if both are subranges of the same type, or if they are identical. Any real variable can have an integer subrange value assigned to it. The conversion from integer to real is automatically done.

At run time, however, the run-time system checks these variables' actual values to see that the range of the variable on the left will allow the value of the variable or expression on the right to be assigned to it. If not, a run-time error occurs.

MULTIDIMENSIONAL ARRAYS

The subrange TYPE declaration facility gives us a very clean, compact way of defining the bounds of matrices and other arrays having more than one dimension. For example, to define a 3 × 3 array, we need only define the type of the subscript and then define the type of matrix in terms of this type. Then we can define all the matrices we might use in terms of the matrix type itself. This works as follows:

```
TYPE
    INDEX=1..3; (*DEFINE RANGE OF INDEX VALUES*)

    (*THEN DEFINE THE MATRIX IN TERMS OF THE INDICES*)

    MATRIX=ARRAY[INDEX,INDEX] OF REAL;
VAR
    X,Y: MATRIX;      (*DEFINES X AND Y
                       AS 3X3 ARRAYS OF REAL*)
```

Pascal also allows the form:

```
X: ARRAY[INDEX1] OF ARRAY[INDEX2] OF REAL;
```

but this is usually more confusing, although it still specifies a two-dimensional array where elements can be referred to as X[I,J].

Pascal also allows whole columns of array data to be referred to in a single statement, regardless of which way the array is specified. In a two-dimensional array X[I,J], the term X[I] refers to the entire *I*th column of the matrix. Thus we could define:

```
TYPE INDEX=1..50;
VAR
    X:ARRAY[INDEX,INDEX] OF REAL (*SQUARE MATRIX*)
    Y:ARRAY[INDEX] OF REAL; (*ONE-DIM ARRAY*)
```

and later in the program move a whole column of X to Y:

```
Y:=X[I];     (*MOVE COLUMN I OF X INTO Y-ARRAY*)
```

The following program allows input of the values of the coupling constants (to be used in a nuclear magnetic resonance calculation program) into a matrix that may be as large as 6×6. The matrix is diagonally symmetrical, so only values to the right of the diagonal are entered. Further, the diagonal is left unspecified since the concept of a nucleus coupled to itself has no physical meaning. Note the use of the subrange types for integer variables as well as for the indices themselves. Note also that the value for the number of spins entered in the procedure GETSPINS must lie in the defined subrange of 1 . . 6, or a run-time error will occur when the program is executed.

```
PROGRAM NMR3;
CONST
    MAXSPINS=6; (*ALLOWS UP TO 6 SPINS*)
TYPE
    INDEX= 1..MAXSPINS; (*INDEX IS A SUBRANGE TYPE*)
    MATRIX= ARRAY[INDEX,INDEX] OF REAL;

VAR
    JVAL:MATRIX;      (*THIS IS A 6 X 6 MATRIX*)
    SPINS: INDEX;     (*CAN TAKE ON VALUES 1..6*)

(********************************************)
PROCEDURE GETJS(N:INDEX);
(*THIS PROCEDURE ALLOWS ENTRY OF THE
MATRIX OF J-VALUES*)
VAR
    K,L:INDEX;   (*BOTH CAN ONLY BE 1 TO 6*)

(*GET JVAL[1,2] TO JVAL[1,N],
THEN JVAL[2,3] TO JVAL [2,N], ETC.*)
BEGIN
    FOR K:= 1 TO N-1 DO (*N,N VALUE IS UNDEFINED AND THUS NOT ENTERED*)
    BEGIN
    FOR L:=K+1 TO N DO (*NOTE L STARTS ABOVE K*)
        BEGIN
        WRITE( 'J[', K:1, ',', L:1, ']= ');
        READLN(JVAL[K,L]);
        JVAL[L,K]:=JVAL[K,L];    (*SYMMETRIC*)
        END; (*FOR L*)
    END; (*FOR K*)
END; (*GETJS*)
(********************************************)
PROCEDURE GETSPINS(VAR N:INDEX);
(*THIS PROCEDURE GETS THE NUMBER
OF SPINS AND RETURNS IT IN N*)
BEGIN
    WRITE('NUMBER OF SPINS= ');
    READLN(N);  (*GET NUMBER OF SPINS*)
END; (*GETSPINS*)
(********************************************)
BEGIN (*MAIN*)
    GETSPINS(SPINS);
    GETJS(SPINS);
(*  :
    :
    :   *)
END.
```

PACKED ARRAYS

Arrays of variables that have limited ranges may sometimes be stored in less memory using *packed arrays.* Suppose that each element of an array could only take on values from 0 to 3. Then it would be possible to store several elements of an array in a single computer word, since numbers between 0 and 3 can be represented in only two bits (00, 01, 10, 11).

In Pascal we do not need to know how numbers are stored or how many bits a computer word contains. Instead, we simply define the subrange of the variable and declare the array to be PACKED. Then the compiler calculates the most efficient packing method for that subrange type. The subrange and packing declarations are shown below for the type SMALLNOS:

```
TYPE
   SMALLNOS=0..3;
VAR
   XNOS:PACKED ARRAY[1..1000] OF SMALLNOS;
```

When we wish to access a particular element of that packed array, we simply refer to it as if it were a normal array:

```
XSUM:=XNOS[3] + XNOS[9];
```

and Pascal will automatically find the correct element. While packing arrays in this manner saves storage, it should be emphasized that it will probably take much more computer time during program execution.

It is not usually possible in most versions of Pascal to read a single element of a packed array:

```
READ(XNOS[5]);  (*USUALLY NOT ALLOWED*)
```

but this can be replaced by:

```
READ(XTEMP);      (*READ INTO SINGLE LOCATION*)
XNOS[5]:=XTEMP;   (*AND THEN PUT
                    INTO PACKED ARRAY*)
```

where XTEMP is also of type SMALLNOS. Single elements of packed arrays can usually be written out, however. Similarly, single elements of packed arrays can be passed to procedures as *value* parameters, but not as reference (VAR) parameters.

PACKING AND UNPACKING

Standard Pascal offers the procedures PACK and UNPACK to convert packed arrays to unpacked ones and vice versa. These procedures can be used to save time when a packed array is accessed continually by unpacking it first into a new array and then, if it has changed, packing it back into the first array when done. However, if there is only one such array in a program, this is a waste of time and space, since you then need room for both arrays instead of for just one.

Suppose we have several packed arrays of SMALLNOS, and we wish to access the elements of one of them many times. We can do this directly by:

```
CONST
     LOWBOUND=1; UPBOUND=1000;
TYPE
     SMALLNOS=0..3;   (*SMALL NUMBERS*)
     ARRAYDIM=LOWBOUND..UPBOUND;

VAR
     I:INTEGER;
     XNOS,YNOS,ZNOS: PACKED ARRAY [ARRAYDIM] OF SMALLNOS;
     XUP: ARRAY[ARRAYDIM] OF SMALLNOS;

BEGIN (*UNPACK XNOS INTO XUP*)
     FOR I:= LOWBOUND TO UPBOUND DO
          XUP[I]:=XNOS[I];
     :
     :
     :
```

Large computer versions of Pascal allow you to abbreviate this FOR loop with the standard procedure UNPACK by writing:

```
UNPACK(XNOS,XUP,LOWBOUND);    (*UNPACK XNOS INTO XUP*)
```

or

```
PACK(XUP,LOWBOUND,XNOS);    (*PACK XUP INTO XNOS*)
```

These are not supported by UCSD Pascal, however.

Note that there is no reason to pack and unpack arrays unless you do not have sufficient storage to keep all arrays unpacked. If you have several arrays which you can pack and have room to unpack the one you are working on, you can save some execution time (at the expense of storage space) in your program.

The program below reads in an array an element at a time, unpacks it completely, and prints out the result.

```
PROGRAM UNPKR;
(*THIS PROGRAM READS IN A SERIES OF NUMBERS
TO A PACKED ARRAY, THEN UNPACKS THEM AND
PRINTS THEM OUT*)

CONST
     RANGE=10;     (*ARRAY DIMENSION*)
TYPE
     SMALLNOS=0..3;   (*SMALL NUMBERS*)
VAR
     I,X:INTEGER;
     XNOS,YNOS,ZNOS: PACKED ARRAY[1..RANGE] OF SMALLNOS;
     XUP: ARRAY[1..RANGE] OF SMALLNOS;

BEGIN
FOR I:=1 TO RANGE DO
(*READ IN THE NUMBERS*)
     BEGIN
     READ(X);     (*GET AS UNPACKED NUMBER*)
     XNOS[I]:=X;      (*PUT INTO PACKED ARRAY*)
     END;
(*UNPACK NUMBERS INTO XUP*)
FOR I:=1 TO RANGE DO
     XUP[I]:=XNOS[I];

(*PRINT OUT THE RESULTS*)
FOR I:=1 TO RANGE DO
     WRITELN('XUP[', I:2, ']= ', XUP[I]);
END.
```

STRINGS AND PACKED ARRAYS OF CHAR

Strings of characters are actually stored within the compiled Pascal program as PACKED ARRAYs OF CHAR. In UCSD Pascal the string has a default maximum length of 80 characters, but a variable actual length, so that any number of characters can be read into the string during execution of the program. The length of the string is then stored in the zeroth element of the string array and can be accessed by:

```
L:= ORD(S1[0]);
```

where L is any integer and S1 is a string. This is not available in DEC-10 Pascal, but Pascal easily allows the construction of such an 80-character packed array to read in strings of characters whose length is not known in advance. The string is then filled to 80 characters with blanks. UCSD Pascal also allows the alphabetic comparison of strings for sorting, and this, too, can be done in DEC-10 Pascal if they are of the same length.

In order to use strings in standard Pascal, we first define the string type:

```
TYPE
    STRING=PACKED ARRAY[1..80] OF CHAR;
```

Then we can refer to any number of string arrays in a single variable definition:

```
VAR
S1,S2,S3:STRING;
```

A routine to read in strings can then be constructed to read characters until an EOLN is found, and then fill the remainder with blanks. The routine GETSTRING is shown in the example in the following section.

READING STRINGS FOR FILENAMES

In UCSD Pascal we can read strings of variable length directly from the terminal, but this is not possible in standard Pascal, and we must devise our own routines to do so. In the program FILER below, the procedure GETSTRING reads characters until a nonblank is found; then it puts the nonblank characters up to the next EOLN into PACKED ARRAY OF CHAR whose dummy name is S and fills with blanks. The procedure skips over any blanks at the beginning in case the file window buffer has a blank in it and since leading blanks would be ignored in sorting strings alphabetically later.

The main program then calls the GETSTRING procedure twice. The first time it gets a string for the file name to use in the REWRITE command. Then the file is opened and a new string is obtained from GETSTRING, which is written into the file. The program then exits and closes the file.

SETS

A *set* is a group of names or objects that go together and are, therefore, of the same type. A particular set may contain some, all, or none of the things of that type. For example, we might declare the type LIQUIDS as being the following list of names:

```
TYPE
    LIQUIDS=(H20,HEXANE,CCl4,CDCl3,CS2,ETOH);
```

```
PROGRAM FILER;
(*THIS PROGRAM READS IN A STRING OF
VARIABLE LENGTH, ASSIGNS 9 CHARACTERS
TO THE OUTPUT FILE NAME, GETS A NEW
STRING AND WRITES IT INTO THAT FILE AND EXITS*)

TYPE STRING=PACKED ARRAY[1..80] OF CHAR;
VAR
     STR:STRING; (*INPUT STRING*)
     FNAME:PACKED ARRAY[1..9]OF CHAR;
     UNPKCHR:CHAR;
     I:INTEGER;
     FVAR:TEXT;   (*TEXT FILE*)
(***********************************)
PROCEDURE GETSTRING(VAR S:STRING);
(*THIS PROCEDURE GETS A STRING FROM
THE TERMINAL. IT IGNORES LEADING SPACES
AND TERMINATES THE STRING AT 80 CHARS*)
VAR K:INTEGER;
     C:CHAR;
BEGIN
REPEAT
     READ(C);
UNTIL C<>' ';    (*SKIP ANY SPACES AT START*)
S[1]:=C;     (*PUT 1ST ONE IN*)
K:=2;    (*AND SET K FOR REST*)
WHILE NOT EOLN AND (K<=80) DO
BEGIN
     READ(C);
     S[K]:=C;     (*PUT IN PACKED ARRAY*)
     K:=K+1;
END;(*WHILE*)
READLN;

     (*FILL REMAINDER OF STRING WITH BLANKS*)
     FOR K:=K TO 80 DO
          S[K]:=' ';
END;     (*GETSTRING*)
(***********************************)
BEGIN   (*MAIN*)
WRITE('FILENAME: ');
GETSTRING(STR); (*READ STRING INTO STR*)

(*COPY 1ST 9 CHARS INTO FNAME*)
FOR I:=1 TO 9 DO
     FNAME[I]:=STR[I];

(*OPEN FILE HAVING THAT NAME*)
REWRITE(FVAR,FNAME);
(*NOW GET STRING FOR FILE*)
WRITE('ENTER STRING FOR FILE: ');
GETSTRING(STR);
(*AND WRITE IT INTO FILE*)
     WRITELN(FVAR,STR);   (*PUT IN FILE*)
END.
```

This is known as an *enumeration* type, and with such a declaration we are only defining a special sort of variable names to the compiler. They are not known at the time the program is executed. However, the use of such names avoids the use of confusing numerical values for each of several solvents we might want to work with, although the compiler actually converts them to numbers at run time. Then variables can be assigned to the type LIQUIDS:

```
VAR SLV:LIQUIDS;
```

and can then take on any of the values from H2O to ETOH. These values can be used as array indices, subrange bounds, and in sets.

Once we have defined these names, we can also declare that there is a type of SET which may contain some, all, or none of these liquids:

```
TYPE
   SOLVS = SET OF LIQUIDS;
```

A set must always be made up of some simple type, such as integer, character, subrange, or some enumeration type such as we showed above. Such types are collectively known as *scalar* types.

Having defined the set of liquids SOLVS, we can now talk about various set variables named POLARSOLVS or NONPOLARSOLVS which have some of the above liquids as members of that set.

```
VAR
   POLARSOLVS, NONPOLARSOLVS:SOLVS;
   (*BOTH ARE SETS OF LIQUIDS*)
```

and can then define the members of each set within the program by simply enclosing the set members in square brackets:

```
POLARSOLVS:= [H2O,CDCl3,CS2,ETOH];
NONPOLARSOLVS:=[HEXANE,CCL4,CS2];
```

Note that *parentheses* are used in defining a type and *square brackets* are used to put members in the set. For the sake of argument we have included CS2 in both the polar and the nonpolar groups.

The main purpose of sets is to allow the compiler to make assignments of numerical values to these names representing related quantities so that they can be compared later.

The names of the members of the set are used only during compilation, and if such names are read later by the compiled program as string variables, they cannot be recognized as set members by the program without programming to compare these strings, such as:

```
GETSTRING(SOLVENT);
IF SOLVENT='CDCL3 ' THEN SLV:=CDCL3;
```

where SLV is a constant of type LIQUIDS and SOLVENT is a string.

SET OPERATORS

The most powerful property of sets in Pascal is the ability to combine and compare sets and check for membership of subrange variables in sets. The operators which Pascal uses for set manipulation are:

−	difference
+	union
*	intersection

=	set equality
< >	set inequality
< =	set contains
> =	set is contained by
IN	set inclusion

Sets can be combined into new sets using the union operator. The *union* of two sets is a new set containing all the elements of both:

```
CHLORSOLVS:=[CCL4,CDCl3];
OXYSOLVS:=[H20,ETOH];
OXANDCHLORSOLVS:= CHLORSOLVS + OXYSOLVS; (*SET
                                            UNION*)
```

The *intersection* of two sets contains the elements common to both sets:

```
POLNONPOL:=POLARSOLVS * NONPOLARSOLVS; (*INTERSEC-
                                           TION*)
```

and in this example, the set POLNONPOL would contain only CS2.

The *difference* of sets contains all the members of the first set which are not members of the second set:

```
NONOXYCHLOR:= SOLVS − OXANDCHLORSOLVS;  (*DIFFERENCE*)
```

Sets can also be compared to see if they are identical, different, contained in, or contained by. The result in each case is the Boolean value TRUE or FALSE.

```
OXYSOLVS <=POLARSOLVS        is thus      TRUE
```

and

```
CHLORSOLVS >=NONPOLARSOLVS   is           FALSE
```

Since we have designated our original list of solvents as a list in parentheses, it is like a subrange type where we use symbols instead of numbers. Such a type is technically called a *scalar* type, and we can test to see whether any scalar value is a member of a set. We use the IN operator to test for set membership. Thus

```
CDCL3 IN CHLORSOLVS      is      TRUE
```

and

```
HEXANE IN POLARSOLVS     is      FALSE
```

Scalar values, once defined, constitute an ordered series, and we can define new subsets from the original list:

```
MODSOLVS:=[HEXANE..CS2];
```

just as we could write

```
LOWALPH:=['A'..'M'];
```

if LOWALPH were a set variable. The above set MODSOLVS then contains HEXANE, CCL4, CDCL3, and CS2. Further we can define empty sets as needed:

```
NEWSOLVS:=[ ];  (*EMPTY SET*)
```

Sets are a useful, compact method of storing data and comparing it to other groups of data. One common use is to retrieve data from a stored list in a file and then compare that data to a set to determine what operation to perform on it. The number of items in a set is implementation dependent, and sometimes it is so small that a SET OF CHAR is not allowed. Subranges of CHAR, such as 'A'..'Z', usually are allowed, however.

In the program below, the strings corresponding to the solvent names are entered using a special version of GETSTRING, which always returns a six-character string, filled with blanks if necessary. These names are then compared with six-character constant strings within the program, and if they match, the scalar variable SLV is set to the corresponding scalar member of the type LIQUIDS. Then membership in the various sets is checked, and if found, a message is printed out. Note the use of the nested IF statements to compare these entered strings with the stored ones. CASE statements cannot be used here since case selectors can only be characters or integers.

```
PROGRAM SOLVID;
(*DEFINE 'LIQUID' AS SCALAR TYPE
CONSISTING OF THE LIQUIDS LISTED.
DEFINE 'SOLV' AS A SET OF LIQUIDS*)
TYPE
     LIQUIDS=(HEXANE,CS2,CCL4,CDCL3,ETOH,H2O,NONE);
     SOLV = SET OF LIQUIDS;
     STRING6=PACKED ARRAY[1..6] OF CHAR;

VAR
(*THE FOLLOWING ARE SETS OF SOLVENTS:    *)
     POLARSOLVS,NONPOLARSOLVS,CHLORSOLVS,OXYSOLVS: SOLV;
     NAME:STRING6;    (*NAME OF ENTERED SOLVENT*)
     SLV: LIQUIDS;    (*CAN TAKE ON VALUE OF ANY ONE SOLVENT*)

(*************************************************)
PROCEDURE GETSTRING (VAR S:STRING6);
(*THIS PROCEDURE GETS A 6-CHAR STRING
ONLY, IT FILLS IF LENGTH IS <6*)
VAR K,L:INTEGER;
     C:CHAR;
BEGIN
REPEAT
     READ(C);     (*READ LEADING SPACES*)
UNTIL C<>' ';
S[1]:=C;          (*PUT IN 1ST NON-BLANK CHAR*)
K:=2;
WHILE NOT EOLN AND (K<=6) DO
     BEGIN        (*GET REST OF CHAR*)
     READ(C);
     S[K]:=C;
     K:=K+1;
     END;(*WHILE*)
FOR L:=K TO 6 DO
     S[L]:=' ';   (*FILL WITH SPACES IF NEEDED*)
READLN;           (*SKIP TO END OF LINE ANYWAY *)
END;
(*************************************************)
```

```
BEGIN
(*PUT MEMBERS IN EACH SET*)
     CHLORSOLVS:=[CCL4..CDCL3];
     OXYSOLVS:=[H2O,ETOH];
     POLARSOLVS:=[H2O,CDCL3,CS2,ETOH];
     NONPOLARSOLVS:=[HEXANE,CCL4,CS2];

(*READ IN A NAME OF UP TO 6 CHARS AND SEE WHAT SETS
IT IS A MEMBER OF*)

REPEAT
     WRITE('ENTER SOLVENT NAME: ');
     GETSTRING(NAME);     (*READ IN NAME*)
     SLV:=NONE;   (*SET IT TO NONE*)

(*COMPARE THE ENTERED NAME AS A STRING E;WITH
EACH OF THE POSSIBLE STRINGS AND
SET 'SLV' EQUAL TO THE CORRESPNDING SCALAR VARIABLE*)

          IF NAME = 'HEXANE' THEN SLV:=HEXANE
ELSE      IF NAME = 'CS2   ' THEN SLV:=CS2
ELSE      IF NAME = 'CCL4  ' THEN SLV:=CCL4
ELSE      IF NAME = 'CDCL3 ' THEN SLV :=CDCL3
ELSE      IF NAME = 'ETOH  ' THEN SLV:=ETOH
ELSE      IF NAME = 'H2O   ' THEN SLV:=H2O;

(* IF ANY OF THE ABOVE WERE FOUND, CHECK FOR SET MEMBERSHIP*)

(*NOTE THAT THESE ARE NOT MUTUALLY EXCLUSIVE
AND THAT SEVERAL MESSAGES ARE POSSIBLE FOR EACH SOLVENT*)

IF SLV <> NONE THEN       (*MUST BE FOUND*)
     BEGIN
     IF SLV IN OXYSOLVS
          THEN WRITELN(NAME, ' CONTAINS OXYGEN');
     IF SLV IN CHLORSOLVS
          THEN WRITELN(NAME,' CONTAINS CHLORINE');
     IF SLV IN POLARSOLVS
          THEN WRITELN(NAME,' IS POLAR');
     IF SLV IN NONPOLARSOLVS
          THEN WRITELN(NAME,' IS NONPOLAR');
     END;(*IF SLV<>NONE*)
UNTIL SLV=NONE; (*QUIT IF NOT A MEMBER OF ANY SET*)
END.
```

A sample execution of this program is as follows:

```
ENTER SOLVENT NAME: ETOH
ETOH   CONTAINS OXYGEN
ETOH   IS POLAR
ENTER SOLVENT NAME: CDCL3
CDCL3  CONTAINS CHLORINE
CDCL3  IS POLAR
ENTER SOLVENT NAME: H2O
H2O    CONTAINS OXYGEN
H2O    IS POLAR
ENTER SOLVENT NAME: NONE

EXIT
```

In assignment statements, sets are compatible in type if their components are of the same base type. At run time a check is made to see that the actual values of one set can fit in the other set.

USING SETS TO RESTRICT THE CASE SELECTOR

The powerful CASE statement discussed in Chapter 5 has the significant limitation that all possible values of the case selector variable must be accounted for in some way. One way of doing this is to restrict the case variable to a subrange type that can only take on a few values, all of which are listed in one case or another.

If, however, several widely separated values of the case variable are to be allowed and all those in between are not allowed, the use of the set to restrict the CASE statement becomes appealing. Let us suppose that we are going to write a program to perform different actions depending on which letter of the alphabet is typed as a command character. Suppose that the only commands will be 'L', 'R', and 'S', representing log, reciprocal, and square root.

We can create a set of alphabetic characters that will be empty except for the three characters allowed, and then check for set membership before allowing the CASE statement to take control. Then only those characters that are members of the set of 'L', 'R', and 'S' will ever be tested by the CASE statement.

To accomplish this, in the declaration part we first declare a set variable made up of alphabetic characters:

```
TYPE
   ALPH= 'A'..'Z'; (*SET OF ALPHABETIC CHARACTERS*)
VAR
   COMD: SET OF ALPH; (*SET OF ALPH CHARS*)
```

and then, in the execution part, we put the three characters into the set:

```
COMD:=['L', 'R', 'S']; (*INITIALIZE SET*)
```

The complete program is shown below. Note that we print an error message if the command is illegal by using an IF-THEN-ELSE statement and another error message if the number entered cannot have its log, reciprocal, or square root taken.

```
PROGRAM CONVERT;
(*THIS PROGRAM CONVERTS THE ENTERED NUMBER TO ITS
NATURAL LOG IF 'L' IS ENTERED,
RECIPROCAL IF 'R' IS ENTERED, AND
SQUARE ROOT IF 'S' IS ENTERED BEFORE THE NUMBER
IT USES THE SET 'COMD' TO CHECK FOR A LEGAL COMMAND*)

TYPE ALPH='A'..'Z'; (*SUBRANGE TYPE OF ALPHABETIC CHARS*)
VAR
C:CHAR;                  (*ENTERED COMMAND CHARACTER*)
COMD: SET OF ALPH;       (*SET VARIABLE*)
ERR:BOOLEAN;             (*ERROR FLAG FOR NEGATIVE OR ZERO ARGUMENTS*)
NUM,RESULT:REAL;         (*INPUT VALUE AND RESULT*)

BEGIN
COMD:=['L','R','S'];        (*INITIALIZE SET*)
                            (*NOTE THAT THIS IS AN EXECUTABLE STATEMENT*)
REPEAT  (*GET VALUES UNTIL A ZERO VALUE FOR NUM IS ENTERED*)
    READLN(C,NUM);          (*GET CHARACTER AND NUMBER*)
    ERR:=FALSE;             (*INITIALIZE TO FALSE*)

    IF C IN COMD THEN       (*CHECK FOR SET MEMBERSHIP*)
        BEGIN
        CASE C OF           (*SEE WHICH ONE IT IS*)
        'L':    IF NUM>0 THEN RESULT:=LN(NUM)
                ELSE ERR:=TRUE;

        'R':    IF NUM<>0 THEN RESULT:=1/NUM
                ELSE ERR:=TRUE;

        'S':    IF NUM>=0 THEN RESULT:=SQRT(NUM)
                ELSE ERR:=TRUE;
        END; (*CASE*)

        (*PRINT ERROR MESSAGE IF ERR IS TRUE
        OTHERWISE PRINT OUT RESULT*)

        IF ERR THEN WRITELN('ILLEGAL VALUE')
        ELSE WRITELN(NUM,' ',RESULT);
        END (*IF C*)

    (*IF NOT A MEMBER OF THE SET, PRINT ILLEGAL COMMAND MESSAGE*)
    ELSE WRITELN( 'ILLEGAL COMMAND');
UNTIL NUM=0.0;  (*EXIT AFTER A ZERO VALUE FOR NUM IS ENTERED*)

END.
```

PROBLEMS

1 Write a program to accept five integers between 1 and 20 and then tell the user whether further integers are members of that set of integers.

2 Write a program to accept words of up to 25 characters and see if those words can be spelled with the letters A, C, D, E, H, I, N, P, R, S, or T.

CHAPTER 11

Records

DEFINITION OF RECORDS

In Pascal a *Record* is a group of data having some relationship, but which may have a mixture of types and values. In a record each member is accessed by name rather than by subscript. Records are, in fact, like arrays at the machine level, but at the Pascal level, we do not need to know how they are structured. We might consider the creation of a group of data regarding some compounds:

Name

Melting point

Boiling point

Solubility in alcohol

Handle under argon (yes or no)

Molecular weight

Nmr spectrum number

In Pascal we can group these data together for each compound in our data by declaring it as a record:

```
TYPE
COMPOUND=
    RECORD
    NAME:STRING;
    MELTINGPOINT, BOILINGPOINT, SOLUBILITY: REAL;
    ARGON: BOOLEAN;
    MW: REAL;
    NMRNUMBER: INTEGER;
    END; (*RECORD*)
```

Note that the record declaration

1. Is a TYPE declaration
2. Begins with RECORD and ends with END
3. May contain all legal types

Once we have declared the RECORD type, we can declare variables of that type, and the Pascal compiler will allocate space for all the components of the record:

```
VAR
   CPD:COMPOUND;   (*CPD IS A RECORD VARIABLE*)
```

Each element of the record variable CPD can be accessed by the record name, followed by a period, followed by the name of the variable within the record:

```
IF CPD.ARGON THEN WRITELN('HANDLE UNDER ARGON!');

IF CPD.SOLUBILITY > 0.5 THEN WRITELN('SOLUBLE IN ALCOHOL');
```

RECORD ARRAYS AND THE WITH STATEMENT

In a useful program the RECORD and ARRAY types can be combined to make a list. Here we define an array of *compounds,* each element of which contains a list of that compound's characteristics:

```
VAR
   CPD:ARRAY [1..RECLENG] OF COMPOUND;
```

If we have to refer to a particular element of this array, we could write:

```
CPD[I].MP:=196.2;   (*SET MELTING POINT*)
CPD[I].NMRNUMBER:= 79;   (*SPECTRUM NUMBER*)
```

If we have to refer to this particular record a large number of times:

```
CPD[I].NAME:='4-ISOPROPYLPYRIDINE';
CPD[I].MP:=-25.6;
CPD[I].BP:=173.0;
CPD[I].SOLB:=1000.0;
    etc.
```

we can abbreviate the continual "CPD[I]." prefix by using the WITH statement.

The WITH statement tells the Pascal compiler that that record is to be referred to implicitly within the statement or compound statement that follows. It has the form:

WITH record identifier DO statement:

The principal purpose of the WITH statement is to save writing the prefix each time. It can also speed up execution, depending on the compiler. We can rewrite the above group of statements as:

```
WITH CPD[I] DO  (*ALL ASSIGNMENTS ARE TO THIS RECORD*)
BEGIN
    NAME:='4-ISOPROPYLPYRIDINE';
    MP:=-25.6;
    BP:=173.0;
    SOLB:=1000.0;
END;
```

However, the WITH statement should be used with care as it may lead to confusion or ambiguity in complex programs.

Records can be treated as single variables in Pascal and moved in a single statement if desired. This will usually gain you some execution speed if the compiler is efficient:

```
VAR C1, C2:COMPOUND; (*RECORDS*)
BEGIN
:
C2:=C1; (*MOVE WHOLE RECORD*)
:
```

RECORDS WITH VARIABLE PORTIONS OR VARIANTS

In some cases it may be desirable to have records whose entries vary for different types of data. For example, for solids we might have:

Name

Molecular weight

Argon handling

Nmr spectrum number

Solubility in alcohol

Melting point

and for liquids we might have:

Name

Molecular weight

Argon handling

Nmr spectrum number

Boiling point

Refractive index

Density

Pascal has gone overboard with records, and allows us to define different record members containing different types of data depending on the value of one scalar variable in the fixed part of the record.

In such records there is a first section called the *fixed part,* which is the same for all records. One of the components of the fixed part must be a variable of a scalar type. The second part of the record is called the *variant part,* and its values depend on what value that scalar type variable takes on.

To declare our record with variants for solids and liquids, we can write:

```
TYPE
     STATE=(SOLID,LIQUID);
     CPD=
     RECORD
          NAME:PACKED ARRAY[1..30] OF CHAR;
          MW: REAL;
          ARGON:  BOOLEAN;
          NMRNO:  INTEGER;
          SOLIQ:  STATE; (*THIS DETERMINES WHICH VARIANT IS USED*)

(*THEN THE TWO VARIANT POSSIBILITIES FOLLOW*)
     CASE STATE OF
          SOLID:(SOLB,MP:REAL);
          LIQUID:(BP,REFRNDX,DENSITY:REAL);
     END;(*CASE AND RECORD*)
```

We see that the variant declaration is also a sort of CASE statement. However, it is not the same in effect. In particular this is part of the declaration section of the program where *types* are selected, rather than blocks of statements. In addition, the list of variables and types selected by the scalar type STATE are enclosed in parentheses. Finally, the END statement that terminates the RECORD statement also terminates the CASE variant section, since the variant section *must* come last in the RECORD definition. Variant parts of records may also be nested, and indeed, records may be made up of records as well, each of which may have variants.

There is also an alternate form for this variant record format in which the variable whose value determines the variant used is part of the CASE statement itself:

```
TYPE
     STATE=(SOLID,LIQUID);
     CPD=
     RECORD
          NAME:PACKED ARRAY[1..30] OF CHAR;
          MW: REAL;
          ARGON:   BOOLEAN;
          NMRNO:   INTEGER;

 (*THE VARIABLE SOLIQ IS DEFINED WITHIN THE CASE STATEMENT*.
     CASE SOLIQ: STATE OF
          SOLID:(SOLB,MP:REAL);
          LIQUID:(BP,REFRNDX,DENSITY:REAL);
     END;(*CASE AND RECORD*)
```

The proposed standard requires that all possible values of the variant selector (like STATE) be accounted for. This usually means that such variables will be a subrange or scalar type.

In reality, the compiler allocates storage for the maximum record size of any of the variants found within the record, and no space is saved by using the smaller variant over a larger variant. Space is saved overall, however, since the various types of variables in the different variants can occupy the same memory. The variant part of a record does provide a sneaky way to access data in two different formats without the compiler issuing a type conflict error message at compile time. (This is actually not allowed according to the proposed standard, but few compilers have implemented this check yet.)

For example, you might want to be able to print out the actual value of a real number or a PACKED ARRAY OF CHAR as if it were an integer. By declaring a record to have a variant part where one variant has a variable of INTEGER type and another variant where the same relative variable has a REAL type, you can put numbers into these locations and move them out again as either type without their being converted from one representation to the other.

FILES AND RECORDS

It is very common to create files made up of records in Pascal. Such files store data much more compactly than text files and can be a mixture of various types of data, depending on the declared record structure. They are simply declared as

```
F1:FILE OF recordname;
```

and can be a series of records of different compounds, students, customers, experiments, or anything else that might be grouped together. For example, we could declare a simple record as:

```
TYPE
   STUD=
   RECORD     (*STUDENT RECORD*)
      ID:INTEGER;        (*ID NUMBER*)
      NAME:STRING;       (*STUDENT NAME*)
      END;
```

and then declare the file of student records as:

```
VAR
   S1:FILE OF STUD;
```

Pascal uses a somewhat different way of reading record variables than a simple characters from a file. This is done through the use of a *record file window,* much like the window in character files. The window pointer is simply a variable which points to a single record area which Pascal automatically reserves for reading in and writing out records. One is reserved for each record file declared. If we declare a file variable name as S1, the record window pointer is named S1↑.

The record into which we have read the data is then accessed through S1↑, and the elements are, in this case, named S1↑.ID and S1↑. NAME.

Then to read the next record into the file window, we give the command:

```
GET (S1);
```

which loads the record into this window. When a file is RESET, the first record is automatically loaded into the window. Thus the file window always contains the *next* record. The function EOF(S1) will become TRUE for the file when an attempt is made to read a record beyond the end of the file. Thus the construction

```
WHILE NOT EOF (file) DO
   BEGIN
   X[I]:=file↑;  (*COPY RECORD INTO ARRAY*)
     :
     :
   GET (file); (*LAST RECORD SETS EOF*)
     :
   END;
```

is generally used.

Writing record values into an output file is equally easy in Pascal. First the values are put in the window for the output file, and then the record is written from the window to the file using the PUT (S1) command. If this file is named T1, the commands are:

```
T1↑.NAME:=ENAME;        (*ENTERED NAME STRING*)
T1↑.ID:=ENUMBER;        (*ENTERED NUMBER*)
PUT (T1);               (*PUT RECORD IN FILE*)
```

Note that while the elements of the record in the record window are referred to by the file variable name followed by an up-arrow, the GET and PUT commands take the file variable name *without* the up-arrow as argument.

It should further be noted that record files like other files are symbolic representations of a magnetic tape and thus can only be read and written sequentially. It is not possible to examine a particular record in a file without reading up to the record with the GET command. Further according to the proposed standard it is never possible to read from a file after writing to it or to write to a file after reading from it without first resetting the file to the beginning. Many Pascals have implementation-dependent extensions for random access of records within files, however. Consult local documentation for details.

FURTHER USES OF GET AND PUT

We noted in Chapter 7 that single characters could be accessed through GET and PUT, although the READ(ch) statement would accomplish more or less the same thing. READ and WRITE statements can only be used with text files, however, in many compilers, and the ability to read and write records requires use of GET and PUT.

In addition, Pascal allows the use of GET and PUT to read and write files of any defined type, such as REAL, INTEGER, or ARRAY OF REAL or INTEGER. For example, to read in data from a file of real numbers, we need only define it as a FILE OF REAL, and then use GET to read in each element:

```
VAR F1:FILE OF REAL;
   X: ARRAY [1..TOP] OF REAL;
BEGIN
   :
   :
I:=1;      (*INITIALIZE INDEX*)
WHILE NOT EOF(F1) DO
   BEGIN
   X[I]:=F1↑; (*PUT INTO ARRAY*)
   GET(F1);      (*GET NEXT ELEMENT INTO WINDOW*)
   I:=I+1;
   END;
   :
END.
```

Similarly, the type could be defined as an array:

```
VAR
   F1:FILE OF ARRAY[1..10] OF INTEGER;
```

and entire 10-number sections read in at once. A file of REAL is used in the FOURIER program in Chapter 19.

It is important to recognize that a FILE OF REAL or FILE OF INTEGER is a much more efficient method of storing data than a text file, since each number takes up only one or two words, while the characters that make up a number may take up many more words in a file.

The proposed standard allows the reading and writing of record files directly with READ and WRITE statements:

```
VAR
ST:STUD;  (*STUDENT RECORD*)
S: FILE OF STUD;  (*FILE OF RECORDS*)
BEGIN
   :
```

READ (S,ST); (*READ ONE RECORD FROM THE FILE*)

This has not yet been implemented in many compilers.

In the example below, the file with the external name STUDENT is given the variable name S1 and is the input file from which student names and ID numbers are read. They are printed out and a grade score is entered. The records are then written, one by one, into a new file with the file variable T1 and the external name NEWSTD which contains the student ID numbers, names, and grades. The file NEWSTD is stored on disk permanently. (In UCSD Pascal the file will only be permanent if the CLOSE(T1,LOCK); statement is given before the end of the program.)

```
PROGRAM GRADER;
(*THIS PROGRAM READS IN RECORDS ONE AT A TIME
AND LISTS OUT THE STUDENT'S NAMES.
IT ALLOWS ENTRY OF GRADES AND STORES THE
RESULTING GRADES AND NAMES IN A NEW FILE*)

TYPE
    STRING=PACKED ARRAY[1..20] OF CHAR;
    STUD=
        RECORD
        ID:STRING;
        NAME:STRING;
        GRADE:INTEGER;
        END;

VAR
    S1, T1: FILE OF STUD;    (*RECORD FILES*)


BEGIN   (*MAIN*)
RESET (S1, 'STUDNT    ');     (*OPEN FILE AND GET 1ST RECORD*)
REWRITE(T1,'NEWSTD    ');     (*OUTPUT FILE*)
REPEAT  (*UNTIL LAST STUDENT HAS BEEN READ AND WRITTEN*)
    WITH S1↑ DO          (*AVOIDS WRITING S1↑ EACH TIME*)
        BEGIN
        WRITE(ID:12,NAME:23);
        READ(GRADE);     (*GET GRADE ENTRY*)
    END;     (*WITH*)
    T1↑:=S1↑;            (*MOVE WHOLE RECORD TO OUTPUT WINDOW*)
    PUT(T1);             (*AND WRITE IT*)
    GET(S1);             (*GET NEXT RECORD*)
UNTIL EOF(S1);         (*QUIT IF LAST ONE READ*)
END.
```

PROBLEMS

1 Write a program to create a series of records of the form:

Compound name
Molecular weight

Melting point
Five principal spectral peaks
and store them in a file.

2 Write a program to read in the previous compound file and print the entries out in alphabetical order.

CHAPTER 12

Sorting

SORTING

There are several ways to sort a list of data in an array into ascending or descending order. The simplest way can be used if there is room for a second array of equal size. Here the smallest number is found in the first array and put into the first element of the second. It is then "crossed out" by setting it to some large number, so that it will not be found again. Then the next smallest number is found, placed in the second element of the second array, and crossed out in the first, and so forth. This simple sort is illustrated below in PROCEDURE SORTX, which sorts the array ARA into the temporary array Y, and then copies it back again sorted into ascending order, where it is printed out:

```
    PROGRAM SORT1;
    (*THIS PROGRAM ALLOWS ENTRY OF 10
    NUMBERS AND THEN SORTS THEM INTO THE NEW ARRAY Y
    BY CROSSING THEM OUT IN THE ARRAY X *)

    CONST
         MAX=1.0E37; (*VERY LARGE NUMBER*)
         TOP=10;
    TYPE
         INDEX=1..TOP;    (*INDEX VARIABLES*)
         XARY=ARRAY[INDEX] OF REAL;
    VAR
         X:XARY; (*ARRAY OF 10 REAL ELEMENTS*)
         I:INDEX;     (*ARRAY INDICES*)
         NXTSML:REAL;

(**********************************************)
PROCEDURE SORTX(VAR ARA:XARY);
VAR Y:XARY; (*LOCAL SCRATCH ARRAY*)
I,J,JSAVE:INTEGER;  (*LOCAL VARIABLES*)
BEGIN
FOR I:=1 TO TOP DO        (*OUTER LOOP FINDS 1 EACH TIME*)
     BEGIN
     NXTSML:=MAX;          (*FIND NEXT SMALLEST*)
     FOR J:= 1 TO TOP DO (*CHECK EACH ONE*)
          IF ARA[J]<NXTSML THEN
               BEGIN       (*SAVE IF SMALLER*)
               NXTSML:=ARA[J];
               JSAVE:=J;   (*SAVE INDEX AS WELL*)
               END;     (*THEN*)
     Y[I]:=NXTSML;        (*SAVE NEXT SMALLEST IN Y-ARRAY*)
     ARA[JSAVE]:=MAX;     (*CROSS OUT THIS X*)
END;(*FOR I*)
(*COPY BACK INTO X *)
     ARA:=Y;       (*IN A SINGLE STATEMENT! *)
END;(*SORTX*)
(**********************************************)
BEGIN (*MAIN PROGRAM*)

(*FIRST GET THE NUMBERS FROM THE KEYBOARD*)
FOR I:=1 TO TOP DO
     BEGIN
     WRITE(I:2, ': ');
```

```
        READLN(X[I]);    (*READ EACH ONE*)
        END;(*FOR I*)
    (*SORT INTO ASCENDING ORDER*)
    SORTX(X);
    (*THEN PRINT OUT SORTED ARRAY*)
    FOR I:=1 TO TOP DO
        WRITELN(I:2,': ', X[I]);
    END.
```

THE BUBBLE SORT

A more efficient sorting routine can be used when there is no room for a duplicate array in the particular computer system. In this method, called the *bubble sort,* the entire array is scanned for the smallest value. This value is then interchanged with that in the first position in the array. Then the elements 2..N of the array are scanned for the smallest value which is then swapped with the value in position 2, and so forth.

Records, as well as simple numerical arrays, can be sorted in this way, allowing sorting by any element of the record. In the program below, 10 student names, identification numbers, and grades are read in from the file STUDIN.DAT and then sorted into alphabetical order, using the bubble sort method. They are then printed out by increasing ID number, using a modified simple sort.

```
PROGRAM STUDENTSORT;
(*PROGRAM TO READ IN A FILE OF STUDENT NAMES, ID NUMBERS
AND GRADES, AND LIST THEM OUT NUMERICALLY AND ALPHABETICALLY*)
CONST
    STMAX=20;
TYPE
    STRING=PACKED ARRAY[1..STMAX] OF CHAR;

STUDENT=
    RECORD
        ID:STRING;
        NAME:STRING;
        GRADE:INTEGER;
    END;

VAR
    S:ARRAY[1..STMAX] OF STUDENT;    (*ARRAY OF RECORDS!!*)
    TMAX,TMIN:STUDENT;   (*RECORD VARIABLES*)
    I,J,ISAVE,MAX:INTEGER;
    IMAX:INTEGER;
    F1:TEXT;
(*******************************************)
PROCEDURE GETSTRING(VAR F:TEXT; VAR S:STRING);
(*THIS PROCEDURE GETS A STRING FROM
THE TERMINAL OR OTHER TEXT FILE. IT IGNORES LEADING SPACES
AND TERMINATES THE STRING AT STMAX CHARS*)
VAR K:INTEGER;
    C:CHAR;
```

```
  BEGIN
  REPEAT
      READ(F,C);
  UNTIL C<>' ';    (*SKIP ANY SPACES AT START*)
  S[1]:=C;     (*PUT 1ST ONE IN*)
  K:=2;    (*AND SET K FOR REST*)
  WHILE NOT (EOLN(F) OR EOF(F)) AND (K<=STMAX) DO
  BEGIN
      READ(F,C);

      S[K]:=C;     (*PUT IN PACKED ARRAY*)
      K:=K+1;
   END;(*WHILE*)
   IF NOT EOF(F) THEN  READLN(F);

   (*FILL REMAINDER OF STRING WITH BLANKS*)
   FOR K:=K TO STMAX DO
       S[K]:=' ';
   END;     (*GETSTRING*)
   (*******************************************)
   (*PROCEDURE TO WRITE OUT THE CURRENT ENTRY*)
   PROCEDURE LISTIT(M:INTEGER);
   BEGIN
       WITH S[M] DO    (*NOTE USE OF WITH TO SIMPLIFY WRITE*)
            WRITELN(ID:15, NAME:25, GRADE:4);
   END;
   (*******************************************)
BEGIN   (*MAIN*)


    RESET(F1,'STUDINDAT');
(*READ IN THE DATA FROM THE INPUT FILE*)
    I:=1;    (*RECORD ARRAY INDEX*)
WRITELN('ORDER READ IN');
WHILE NOT EOF(F1) DO
    BEGIN
    GETSTRING(F1,S[I].ID);
    IF NOT EOF(F1) THEN GETSTRING(F1,S[I].NAME);
    IF NOT EOF(F1) THEN READLN(F1,S[I].GRADE);
    IF NOT EOF(F1) THEN
         BEGIN
         LISTIT(I);  (*LIST OUT AS READ IN*)
         I:=I+1;
         END;
    END; (*WHILE*)
MAX :=I-1;  (*TOTAL READ IN*)
(*ALPHABETIZE ENTRIES IN PLACE BY BUBBLE SORT*)
WRITELN;
WRITELN('ALPHABETICAL ORDER');

FOR I:=1 TO MAX DO
    FOR J:=I TO MAX DO
    IF(S[I].NAME>S[J].NAME) THEN
         BEGIN
         TMAX:=S[I]; (*SAVE LARGEST NAME*)
         S[I]:=S[J]; (*SWAP WHOLE RECORD EACH TIME!! *)
         S[J]:=TMAX;
         END;(*IF*)

(*WRITE OUT THE ALPHABETIZED ARRAY AND ID NUMBERS*)
FOR I:=1 TO MAX DO
    LISTIT(I);

(*PRINT OUT IN ASCENDING ORDER BY ID NUMBER*)
(*HERE WE JUST LOOK FOR THE LEAST ONE GREATER THAN TMIN*)
```

```
  TMIN.ID:='000000000
  WRITELN;
  WRITELN('NUMERICAL ORDER');

  FOR J:=1 TO MAX DO
      BEGIN
      ISAVE:=1;
      TMAX.ID:-'999999999               ';(*BECOMES SMALLER EACH TIME*)
      (*SCAN THROUGH ARRAY LOOKING FOR NEXT GREATEST VALUE*)
      FOR I:=1 TO MAX DO
          IF (S[I].ID<TMAX.ID)
              AND (S[I].ID>TMIN.ID) THEN
              BEGIN
              ISAVE:=I;
              TMAX.ID:=S[I].ID;
              END; (*IF S[I] *)
      LISTIT(ISAVE);  (*WRITE OUT THAT ONE EACH TIME*)
      TMIN:=TMAX; (*NOW LOOK FOR NEXT GREATEST*)
      END;(*FOR J*)
END.
```

SORTING BY THE SHELL INTERCHANGE SORT

One of the most efficient sorting methods is the so-called *Shell* interchange sort, named after its inventor, Donald Shell. In this sorting method the array is sorted by interchanging pairs of numbers some distance apart as long as the larger one is first. Then the distance is decreased and the array scanned again. This is repeated until the distance has decreased to 1 and no interchanges have occurred during a scan through the array.

In a 10-element array, for example, the elements 1 and 6 are tested and swapped if element 6 is greater than 1, then 2 and 7 are tested, and then 3 and 8, etc. Then the array is scanned repeatedly using this same distance until no interchanges occur. Then the distance is decreased by a factor of 2. In this case 5 DIV 2 − 2, and the array is scanned repeatedly using this new distance until no interchanges occur. Finally the array is scanned with an array element separation of 1 until no interchanges occur.

To design a Pascal program for this sorting method, we can use the Boolean variable SWAP to indicate if an interchange occurred, and the integer SEP to indicate the array element separation. Our outer loop will be:

```
REPEAT
  :
  :
UNTIL SEP=1;
```

The next inner loop will scan the array for a given distance until no swaps occur:

```
REPEAT

UNTIL NOT SWAP; (*QUIT WHEN FALSE*)
```

In the innermost loop the array is scanned repeatedly:

```
FOR I:=1 TO (TOP-SEP) DO
   IF X[I] > X[I+SEP] THEN
   swap elements and set SWAP to TRUE.
```

```
PROGRAM SHELL;
(*THIS IS THE  PASCAL VERSION OF THE OLD SHELL GAME:
HERE AN ARRAY OF 10 REAL NUMBERS IS READ IN
AND SORTED BY THE SHELL SORTING PROCEDURE*)

CONST TOP=10;

TYPE INDEX=1..TOP;
VAR
     X:ARRAY[INDEX] OF REAL; (*ARRAY TO BE SORTED*)
     I,SEP:INTEGER;   (*INDEX AND SEPARATION*)
     XSAVE:REAL; (*X SAVED DURING SWAP*)
     SWAP:BOOLEAN;    (*SET TO TRUE IF SWAP OCCURS*)

BEGIN
(*FIRST READ IN THE ARRAY*)
FOR I:=1 TO TOP DO
     BEGIN
     WRITE('X[', I:2, ']=');
     READLN(X[I]);
     END; (*FOR I *)

(*START WITH SEPARATION AS LARGE AS ARRAY AND CUT
IT IN HALF AT THE START OF EACH PASS*)
SEP:=TOP;

REPEAT
SEP:=SEP DIV 2; (*CUT SEPARATION IN HALF FOR EACH PASS*)
     REPEAT (*UNTIL NO CHANGES AT EACH SEP*)
     SWAP:=FALSE; (*INITIALIZE TO FALSE*)
     FOR I:=1 TO (TOP-SEP) DO
          BEGIN
          IF X[I]>X[I+SEP] THEN
          (*SWAP 2 ELEMENTS IF LOWER IS GREATER*)
               BEGIN
               XSAVE:=X[I];
               X[I]:=X[I+SEP];
               X[I+SEP]:=XSAVE;
               SWAP:=TRUE; (*SET TO TRUE IF A SWAP IS DONE*)
               END;(*IF*)
          END;(*FOR*)
     UNTIL NOT SWAP;
UNTIL SEP=1;

(*WRITE OUT THE SORTED ARRAY*)
FOR I:=1 TO TOP DO
     WRITELN('X[', I:2, ']=', X[I]);
END.
```

CHAPTER 13

The Pointer Type

USING POINTERS TO ALLOCATE STORAGE

Pascal provides a relatively unique facility for allowing values to point to each other through the use of the *pointer* type. A variable of this type simply contains the memory address where the variable pointed to is stored. The type of data pointed to is defined by declaring at the beginning of the program that the pointer points to that type:

```
VAR IP: ↑INTEGER; (*IP POINTS TO VAR OF TYPE INTEGER*)
```

That pointer variable IP is then said to be *bound* to the variable type INTEGER. Pointers can take on values pointing to any type of data structure or may point to nothing at all. In this latter case they are said to have been assigned the value NIL. Thus the value NIL is a *reserved word* in Pascal, corresponding to some special number for each computer system which is *not* a valid data address.

Pointers cannot point to variables to which we can already refer directly, but must point to variables, arrays, or other structures that have been created by the NEW command.

One of the simplest and most useful purposes to which the pointer type can be put is allocating space for temporary arrays which are later to be dismissed so that the memory space can be reused for some other kind of data. This can be done by declaring the array to be created as a type and then declaring a pointer to that type:

```
TYPE
   XA=ARRAY[1..1000] OF REAL;
VAR
   XP:↑XA;      (*XP IS A POINTER TO THAT ARRAY*)
```

Then we simply allocate space during the program with the NEW statement:

```
NEW (XP);   (*ALLOCATE SPACE
            FOR ONE ARRAY OF TYPE XA*)
```

Such storage space is allocated from a section of free memory referred to as the *heap.* The value of such a variable or the contents of such an array are initially undefined and must have values assigned to them before they can be used.

The array is then referred to through the pointer as:

```
XP↑[I]
```

where this symbolism means that XP is the pointer to the data and that XP↑ is the data pointed to. Individual elements of that array of type XA are then addressed as XP↑[I] and can be treated as array elements in the usual sense.

In standard Pascal this memory can then be disposed of for reallocation to data of a different type with the DISPOSE command:

```
DISPOSE(XP); (*RELEASE MEMORY ALLOCATED BY NEW*)
```

In summary:

1 We declare a pointer type by *preceding* the name of the type to which it points with an up-arrow in either a TYPE or a VAR declaration:

```
TYPE
 PTR=↑INTEGER;    (*DEFINE PTR TO INTEGER TYPE*)
 XARY=ARRAY[1..1000] OF REAL;
VAR
 R,P: ↑REAL;      (*R AND P POINT TO REAL VARIABLES*)
 X:↑XARY;         (*X IS A PTR TO A VAR OF TYPE XARY*)
 K:PTR;      (*K IS A PTR TO A VAR OF TYPE INTEGER*)
 V:REAL;
```

2 We allocate more storage of either single elements or structures with the NEW (ptr) command, where the argument to NEW is a pointer to the element or structure to be allocated:

```
NEW(P);
```

3 The value of a pointer such as P can be copied to another pointer. The up-arrow is not used:

```
R:=P;
```

4 The value of the data pointed to is referred to by *following* the pointer with an up-arrow:

```
V:=R↑;   (*GET REAL VALUE*)
```

5 An element of an array allocated by a NEW command is referred to by giving the pointer name followed by an up-arrow followed by the index in brackets:

```
NEW(X);        (*ALLOCATE SPACE FOR
                1000-ELEMENT REAL ARRAY ON HEAP*)
X↑[I]:=V;      (*PUT VALUE IN ONE ELEMENT
                OF NEW ARRAY*)
```

6 Any data space allocated can be deallocated so that it can be reused with the DISPOSE(ptr) statement:

```
DISPOSE(P); (*DE-ALLOCATE ONE REAL ELEMENT*)
DISPOSE(X); (*DE-ALLOCATE ONE REAL ARRAY*)
```

THE LINKED LIST

One of the most powerful features of the pointer type in Pascal is that of constructing linked lists, where the list can be made up of any type of data. These data can then be linked together by some relationship such as alphabetical order or numerical order.

A linked list is one that contains both data and pointers to other data, and thus each element of that list must have both character or numerical values and pointer(s) to other such list elements. This structure can only be accomplished using a RECORD declaration to define both the data and the pointers to other data:

```
TYPE
INDIV=
    RECORD              (*RECORD ELEMENT OF LINKED LIST*)
    NAME:STRING;        (*SOME PACKED ARRAY OF CHARACTERS*)
    SCORE:INTEGER;      (*SOME SCORE VALUE*)
    NEXT:↑INDIV;        (*POINTER TO ANOTHER RECORD*)
    NAMPNT:↑INDIV;      (*ALPHABETICAL PTR TO NEXT RECORD*)
    CROSSOUT:BOOLEAN;      (*SET TO TRUE WHEN XED OUT*)
    END;      (*RECORD*)

VAR

    IREC:↑INDIV;        (*IREC IS A POINTER TO SUCH A RECORD*)
```

In a program, in order to read in a list of records and connect them with these pointers, we create a new structure each time with the NEW command:

```
NEW(IREC); (*GET ANOTHER RECORD SPACE*)
```

and read in the data to be put in this record. The pointer to the record created by NEW(IREC) is then named IREC and the record itself accessed by IREC↑. Individual elements of this record are named IREC↑. NAME, for instance, and can be fetched by the general string input routine GETSTRING as follows:

```
GETSTRING(IREC↑.NAME);
```

LINKING SUCCESSIVE RECORDS TOGETHER

In order for us to be able to access each of these records we must then construct a link from the newest one to the previous one and keep a pointer that will allow us to find the whole list, an element at a time. We do this by creating a pointer, often named BASE, which points to the last created record. We sometimes say that our list is *anchored* to the pointer BASE. Then we make the newest record point to the one before that, and so forth, so that the entire list is linked together up to the last element which points to NIL. In other words, BASE points to the most recent element, each middle element points to the previous one, and the earliest element points to NIL.

We thus start reading in a series of data from a linked list by setting the pointer BASE equal to NIL:

```
VAR IREC, BASE:↑INDIV;   (*BOTH ARE PTRS TO RECORD
                           INDIV*)
BEGIN
BASE:=NIL; (*SET BASE POINTER TO NIL INITIALLY*)
```

Then we call for Pascal to allocate one record of storage using the NEW command:

```
NEW(IREC); (*GET ANOTHER RECORD FROM THE HEAP*)
```

put data in this new record:

```
GETSTRING(IREC↑.NAME); (*READ IN NAME*)
```

and insert it in the linked list by setting the pointer IREC↑.NEXT to the current value of BASE, so that this NEXT pointer points to the old value of BASE, and setting BASE to point to the current value of the pointer IREC:

```
IREC↑.NEXT:=BASE;   (*PT TO OLD VALUE OF BASE*)
BASE:=IREC;   (*AND PT BASE TO CURRENT RECORD*)
```

This looks like this schematically:

Structure **Statements**

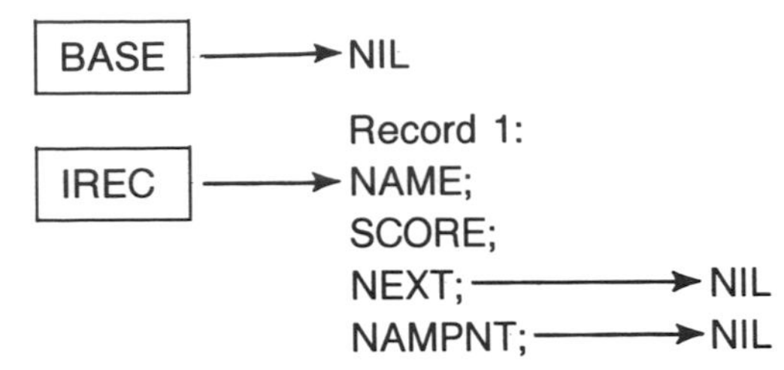

BASE:=NIL;

NEW(IREC);

IREC↑.NEXT:=NIL;
IREC↑.NAMPNT:=NIL;

After the assignment statements:

```
IREC↑.NEXT:=BASE;
BASE:=IREC;
```

we have the situation:

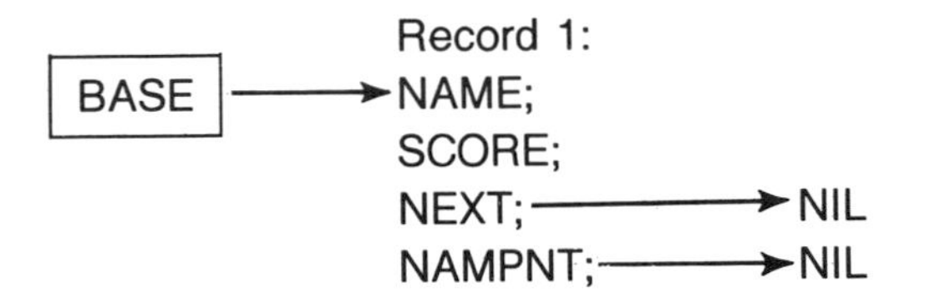

In the second cycle through this loop, we again execute the same two assignment statements to add a new record and see that BASE now points to *this* second record and that *its* NEXT pointer then points to the first record:

```
IREC↑.NEXT:=BASE;   (*NEXT PTS TO OLD BASE*)
BASE:=IREC;   (*CHANGE BASE TO PT TO NEWEST RECORD*)
```

```
                Record 2:
| BASE | ----> NAME;
                SCORE;                  Record 1:
                NEXT; ----------------> NAME;
                NAMPNT; --> NIL         SCORE;
                                        NEXT; ----> NIL
                                        NAMPNT; --> NIL
```

When this process is complete, BASE will point to the latest record allocated, and each record NEXT pointer will point to the most recent previous record allocated up to the first one which will have a NEXT pointer pointing to NIL.

SCANNING THROUGH THE LINKED LIST

One way of referring to a few records is by combining sequences of pointer types, so that the name of the next previous record is obtained. For example, if the name in the most recent record is:

BASE↑.NAME

then the name in the one before this is:

BASE↑.NEXT↑.NAME

and the name in the one linked to this is:

BASE↑.NEXT↑.NEXT↑.NAME

and so forth.

A more efficient way of moving through this list a record at a time is to start with a pointer equal to BASE and advance it each time by setting the new pointer equal to the one pointed to by the NEXT pointer. We repeat this until the NEXT pointer is NIL:

```
IREC:=BASE; (*SET IREC POINTER EQUAL TO BASE POINTER*)
REPEAT
     WRITELN(IREC↑.NAME);      (*WRITE OUT EACH NAME*)
     IREC:=IREC↑.NEXT;         (*SET PTR TO NEXT RECORD*)
UNTIL IREC=NIL;              (*UNTIL THERE ARE NO MORE*)
```

In the program RECSORT given below, we read in student names from the terminal or INPUT file as text and then print them out and ask for a grade which is stored in the record. The records are tied together by the pointer .NEXT, which points to the next previous one allocated. Then the records are sorted in place without being moved by finding the largest one and setting a base pointer NBASE to it. Then the Boolean CROSSOUT is set in this record so that it will not be found again, and the next largest name that is not "crossed out" is found. The pointer NBASE is set to point to it, and its .NAMPNT pointer is set to point to the previously found name (which NBASE points to), and so forth, until all names have been alphabetized. At the end of this sort there are N records which are linked in two ways:

1 By order of entry with BASE pointing to the last one entered, and
2 By alphabetical order, with NBASE pointing to the lowest one in alphabetical order

Naturally it would also be possible to link them in further ways, such as by ID number or score if desired, all without moving any entries in the list of records.

```
PROGRAM RECSORT;
(*PROGRAM TO READ IN NAMES FROM THE
INPUT FILE OR TERMINAL, PUT THEM IN RECORDS, AND
ALPHABETIZE THEM BY POINTERS*)
CONST
     STMAX=30;    (*MAX STRING LENGTH*)
TYPE
     STIND=1..STMAX;
     STRING=PACKED ARRAY[STIND] OF CHAR;
     INDIV=
          RECORD
          NAME:STRING;
          SCORE:INTEGER;
          NAMPNT:↑INDIV;    (*PTR TO NEXT ALPHABETICAL NAME*)
          NEXT:↑INDIV;      (*PTR TO NEXT IN LINKED LIST*)
          CROSSOUT:BOOLEAN;    (*TRUE IF CROSSED OUT*)
          END; (*RECORD DEFN*)
VAR
     BASE,NBASE,SBASE,IREC,PTRSAV:↑INDIV;
     I,J,K,N:INTEGER;
     NAMEMAX:STRING;
     F1:TEXT;      (*INPUT DATA FILE OF NAMES AND SCORES*)

(****************************************************)
PROCEDURE GETSTRING(VAR F:TEXT; VAR S:STRING);
(*THIS PROCEDURE HAS BEEN SHOWN EARLIER
IN THE 'STUDENTSORT' PROGRAM. *)
END;     (*GETSTRING*)
(****************************************************)


BEGIN (*MAIN*)
RESET(F1,'SORTIN   ');  (*OPEN 'SORTIN' FOR INPUT*)
N:=0;          (*COUNTS RECORDS READ IN*)
BASE:=NIL;   (*SET BASE POINTER INITIALLY TO NIL*)
REPEAT         (*UNTIL FILE IS EMPTY*)
     NEW(IREC);   (*ACQUIRE A NEW RECORD ON THE HEAP*)
     GETSTRING(F1,IREC↑.NAME);    (*READ IN NAME FROM INPUT FILE*)
     (*PRINT IT ON THE TTY AND GET SCORE*)
     WITH IREC↑ DO
```

```
        BEGIN
        WRITE(NAME);
        READLN(SCORE);  (*READ IN THE SCORE FROM TERMINAL*)
        N:=N+1;                  (*COUNT RECORDS READ IN*)
        NAMPNT:=NIL;        (*INITIALIZE PTR TO NIL*)
        NEXT:=BASE;      (*MOVE PTR ON TO NEXT*)
        CROSSOUT:=FALSE;    (*INIT CROSSED OUT STATE*)
        END;(*WITH*)
        BASE:=IREC;              (*KEEP BASE POINTING TO THIS ONE*)
UNTIL EOF(F1);  (*REPEAT UNTIL NO MORE IN INPUT FILE*)

(*ALPHABETIZE THE DATA BY SEARCHING THROUGH THE RECORDS
TO FIND THE SMALLEST NAME NOT ALREADY ASSIGNED A POINTER TO A
'NEXT' NAME AND POINTING NBASE TO IT. *)

NBASE:=NIL;            (*START NAME BASE AT NIL*)
FOR I:= 1 TO N DO     (*REPEAT N TIMES TO GET ALL OF THEM*)
     BEGIN
     FOR K:= 1 TO STMAX DO NAMEMAX[K]:='A';  (*NAME IS ALL A'S*)
     IREC:=BASE;        (*START AT BASE OF PTRS FOR RECORDS*)
     REPEAT             (*UNTIL ALL CHECKED AND NEXT IS NIL*)
         IF (IREC↑.NAME > NAMEMAX) AND
           (NOT IREC↑.CROSSOUT) THEN
             BEGIN
             NAMEMAX:=IREC↑.NAME;      (*SAVE THIS NAME*)
             PTRSAV:=IREC;             (*AND THE POINTER*)
             END;     (*IF*)
         IREC:=IREC↑.NEXT;             (*MOVE ON TO NEXT RECORD*)
     UNTIL (IREC=NIL);                 (*KEEP IT UP UNTIL NO MORE LEFT*)

(*THEN SET NEXT NAME EQUAL TO SMALLEST ONE FOUND*)
     PTRSAV↑.NAMPNT:=NBASE;
     NBASE:=PTRSAV;
     PTRSAV↑.CROSSOUT:=TRUE;  (*CROSS OUT AS ALPHABETIZED*)
     END;     (*FOR*)

(*PRINT OUT THE ALPHABETIZED LIST*)
     IREC:=NBASE;              (*START AT TOP OF LIST*)
     REPEAT
         WITH IREC↑ DO
         BEGIN
         WRITELN(NAME,SCORE:5);
         IREC:=NAMPNT;    (*GO ON TO NEXT NAME*)
         END;(*WITH*)
     UNTIL IREC=NIL;
     END.
```

INSERTING RECORDS IN A LINKED LIST

The general procedure for inserting records in a linked list is very simple:

1. Find the element just before the position you wish to insert.
2. Set the new element's pointer to the one the just preceding element originally pointed to.
3. Set the NEXT pointer of the element before the insert position to point to the new element.

This has the general form:

```
(*IREC IS POINTING TO THE ELEMENT BEFORE THE INSERT POSN*)
(*NEWP↑ POINTS TO NEW RECORD TO INSERT*)

NEWP↑.NEXT:=IREC↑.NEXT; (*SET NEW ELEMENT TO PTR OF OLD ONE*)
IREC↑.NEXT:=NEWP;        (*SET POINTER TO PT TO NEW ELEMENT*)
```

Inserting an element *before* an element in a list is more difficult, since we cannot find out directly which element is pointing to it. So, instead, we remember the pointer to that element and search for it in the list of NEXT pointers:

```
(*IREC IS POINTING TO THE ELEMENT BEFORE WHICH WE WISH TO INSERT*)
  PTRSAV:=IREC;    (*REMEMBER THIS POINTER*)

  (*THEN SCAN THROUGH THE LIST TO FIND A NEXT POINTER EQUAL TO IT*)
  IREC:=BASE;      (*SET POINTER TO BEGINNING OF LIST*)

  WHILE IREC<>PTRSAV DO
          IREC:=IREC↑.NEXT;
```

Then we simply insert the pointer to the new element there, as before:

```
NEWP↑.NEXT:=IREC↑.NEXT
IREC↑.NEXT:=NEWP;
```

OTHER METHODS OF SPECIFYING RECORD POINTERS

While it is usual in creating a linked list to have only one pointer, it is possible to have pointers to both the next and the previous elements to allow immediate access to the list in either direction. This leads, of course, to more storage per record, but it prevents having to scan for the correct pointer as shown above. Such a list is called *doubly linked.* In addition, the two end elements can be pointed to each other so that there is no beginning or end, but a circle of elements, although, of course, the BASE pointer must point to something.

On occasion we might also have to refer to two connected elements of a linked list. This can be done by stringing the record pointers together as follows:

IREC	refers to the pointer to the record
IREC↑	refers to the contents of that record
IREC↑.NAME	refers to the name entry pointed to by IREC
IREC↑.NAMPNT	refers to the pointer to the next record
IREC↑.NAMPNT↑.NAME	refers to the next name in the name list

PROBLEMS

1 Write a program to read in 10 integers and then sort them into ascending order without moving them.

2 Expand the above program so that for any entered number after the first 10, the program will print out the number on either side and then enter the new one in the proper sorted place in the list. Stop after 20 entries or when zero is entered.

CHAPTER 14

The GOTO Statement

One of the least structured "last resort" statements available in some versions of Pascal is the GOTO statement. As you can see, it is seldom used, even in complex programs, since we are introducing it last. There is a strong feeling among many programmers that statements that allow direct transfer of control to other parts of the program can lead to a very confusing, unreadable code, which is thus "unstructured." For this reason UCSD Pascal actually prevents use of the GOTO statement unless the compiler comment

```
(*$G+*)
```

is set. This is not necessary in DEC-10 Pascal. The purpose of this restriction is to discourage novice programmers from using it unnecessarily. This restriction is often also enforced in programming courses.

The GOTO statement allows direct transfer of control to another line in the program where a particular destination statement has been *labeled.* In Pascal each label is an integer between 0 and 9999 which must be declared as a label and used as a label by following it with a colon, similar to CASE statement selectors. The label declaration format is

```
LABEL
   1,2,45,99;
```

and it must come before any other declarations. Thus it appears first, even before the CONST declaration. The order of declarations is then:

```
LABEL 4,55,7;
CONST STRLENG=80;
TYPE STRING=PACKED ARRAY[1..STRLENG] OF CHAR;
VAR A,B,X:REAL;
   S1:STRING;
```

When a label is used in a GOTO statement, the statement is written:

```
GOTO 5; (*TRANSFER CONTROL TO LABELED STATEMENT 5*)
 :
 :
5: READLN(A); (*CONTROL IS TRANSFERRED HERE*)
```

Obviously, each label can appear only once.

The principal objection to the use of the GOTO statement is that it can lead to some rather confusing jumping around if it is used promiscuously, and that there are always more structured statements available for accomplishing the same thing.

For example, one might write:

```
REPEAT
    READ(C);
    S[I]:=C;
    I:=I+1;
    IF EOF THEN GOTO 10;
UNTIL I>80;
10: WRITELN('LOOP IS DONE');
```

but one can more appropriately write:

```
REPEAT
    READ(C)
    S[I]:=C;
    I:=I+1;
UNTIL (I>80) OR EOF;
WRITELN('LOOP IS DONE');
```

A major use of the GOTO statement is the transfer of control out of deeply nested loops when unexpected data prevents further calculations. This can always be done by setting some Boolean variable to FALSE if the data is bad and exiting on that Boolean from a WHILE or REPEAT loop, but this is sometimes a bit of a circumlocution.

One of the problems of illustrating the GOTO statement is that cases where it might reasonably be used are never simple, and thus the illustrated uses seem silly at first. Thus we give below a simple example of the GOTO statement, but call your attention to the MATINV program in Chapter 17, where comments indicate how the program could have been written using GOTO statements to exit from an inner loop when the matrix is found to be singular.

Let us consider a simple example of a program that calculates the product of the elements in two similar arrays and then divides all elements in one of the arrays to date by that product. We will not concern ourselves either with how the data got into the arrays or why anyone would want to carry out such an operation. We could write such a program as follows:

```
PROGRAM XYDIV;
(*DIVIDES ALL ELEMENTS UP TO I OF THE X-ARRAY
BY THE PRODUCT OF THE CURRENT X AND Y VALUE
EXITS IF A ZERO DENOMINATOR "PROD" IS FOUND *)

CONST TOP=10;    (*ARRAY DIMENSION*)
VAR
    X,Y:ARRAY[1..TOP] OF REAL;
    I,J:INTEGER;
    PROD:REAL;
    QUIT:BOOLEAN;    (*INDICATES THAT DIVISION BY ZERO WOULD OCCUR*)

BEGIN
(*  :
NEVER MIND HOW THE DATA GETS INTO THE ARRAYS
    :   *)
QUIT:=FALSE;          (*FLAG TO STOP SET TO FALSE INITIALLY*)
I:=1;               (*INITIALIZE LOOP VARIABLE*)
REPEAT
    PROD:=X[I]*Y[I];       (*CALC PRODUCT*)
    QUIT:= PROD=0;         (*TRUE IF ZERO PRODUCT*)
    IF NOT QUIT THEN
        FOR J:=1 TO I DO
            X[J]:=X[J]/PROD;     (*DIVD IF NOT 0 *)
UNTIL QUIT OR (I>TOP);   (*QUIT IF DIV BY 0 OR IF ALL I'S DONE*)
END.
```

Now we can simplify the QUIT construction using the GOTO by simply jumping out of the outer loop if any PROD becomes zero and thus exiting from the program at once. A more complete program might generate an error message in such a case. Note that the LABEL 99 statement must come even before the CONST section and that the QUIT Boolean is totally eliminated by this program:

```
PROGRAM XYDIV2;
(*DIVIDES ALL ELEMENTS UP TO I OF THE X-ARRAY
BY THE PRODUCT OF THE CURRENT X AND Y VALUE
EXITS IF A ZERO DENOMINATOR "PROD" IS FOUND *)

LABEL 99;         (*LABEL DEFINED HERE, MUST BE BEFORE CONST *)
CONST TOP=10;     (*ARRAY DIMENSION*)
VAR
    X,Y:ARRAY[1..TOP] OF REAL;
    I,J:INTEGER;
    PROD:REAL;

BEGIN
(*  :
NEVER MIND HOW THE DATA GETS INTO THE ARRAYS
    :   *)
FOR I:=1 TO TOP DO
    BEGIN
    PROD:=X[I]*Y[I];       (*CALC PRODUCT*)
    IF PROD=0 THEN GOTO 99; (*SKIP OUT IF DIVN BY 0 WILL OCCUR*)
    FOR J:=1 TO I DO
        X[J]:=X[J]/PROD;     (*DIVD IF NOT 0 *)
    END;(*FOR*)
99:  END.
```

While this program is indeed simpler, it is not quite so readable, since you can arrive at the final END. statement unexpectedly through the GOTO rather than linearly as before. Further, transfer out of the middle of the loop is itself unexpected and is not really documented by the form of the FOR loop itself.

The proposed standard requires that the GOTO label be no more distant than the outermost level of nesting of the current procedure. A few compilers allow the label to be anywhere in the program.

Note in summary that the major declarations at the beginning of the program must now be:

```
PROGRAM name;  (*optional*)
LABEL list;  (*optional*)
CONST list;  (*wise*)
TYPE list;  (*useful*)
VAR list;  (*always required *)
```

CHAPTER 15

Debugging a Pascal Program

DEBUGGING

The Pascal language is designed so that many common programming mistakes are precluded by the structure of the language. Thus most of the errors in Pascal programs are fairly easy to find by simply having the program print out intermediate results. In this chapter we will take up a buggy program to calculate the "well-known" Gronch constant, which is simply the sum of the odd elements of an array divided by twice the sum of the even elements of the array:

$$\text{GRONCH} = \frac{\text{ODDSUM}}{\text{EVENSUM} \times 2}$$

In order to test any program, it is essential that we have a suitable hand-calculated case so that we can tell whether we have obtained the correct answer. For this test we will create an input test file of the form

1.0
2.0
3.0
:
.
9.0
10.0

In this array the Gronch constant can be calculated by:

$$\frac{(1.0 + 3.0 + 5.0 + 7.0 + 9.0)}{2(2.0 + 4.0 + 6.0 + 8.0 + 10.0)} = \frac{25}{30}$$
$$= 0.41667$$

Below we present the first version of our buggy program with the line numbers from a typical listing included for easy reference during our discussion:

```
 100    PROGRAM BUGGY1;
 200    (*THIS PROGRAM CALCULATES THE GRONCH CONSTANT
 300    WHICH IS EQUAL TO THE SUM OF THE ODD ELEMENTS
 400    DIVIDED BY TWICE THE SUM OF THE EVEN ELEMENTS OF ANY ARRAY*)
 500
 600    CONST
 700            MAX=10; (*ARRAY MAXIMUM*)
 800
 900    VAR
1000            X: ARRAY[1..MAX] OF REAL;         (*ARRAY TO GRONCH ON*)
1100            N,I:INTEGER;      (*MAX IN ARRAY AND INDEX*)
1200            ODDSUM,EVENSUM,GRONCH:REAL;
1300            F1:TEXT;          (*INPUT TEXT FILE*)
1400
```

```
1500    BEGIN
1600            RESET(F1,'GRONCH.DAT');
1700    (*READ IN DATA UNTIL END OF FILE*)
1800    REPEAT
1900            READLN(F1,X[I]);
2000            I:=I+1;
2100    UNTIL EOF(F1);
2200    (*REMEMBER NUMBER READ IN*)
2300    N:=I;
2400
2500    (*DO ODDSUM*)
2600    REPEAT
2700            ODDSUM:=ODDSUM+X[I];
2800            I:=I+2;
2900    UNTIL I=N;
3000    (*NOW DO EVEN SUM*)
3100    REPEAT
3200            EVENSUM:=EVENSUM+X[I];
3300            I:=I+2;
3400    UNTIL I=N;
3500
3600    (*NOW CALCULATE THE GRONCH CONSTANT*)
3700    GRONCH:=ODDSUM/EVENSUM*2;
3800    WRITELN('GRONCH CONSTANT= ',GRONCH);
3900    END.
```

When we run this program the first time, we get the error message

? ARRAY INDEX OUT OF BOUNDS

or some equivalent statement along with some other numerical information that may mean little or nothing, depending on the computer system in use. Obviously we have made some mistake in the array index calculation, so we look at the lines around the BEGIN statement where the array X[I] is first referred to. Looking at lines 1500–2000, we see that the array index I is never set to any value, and thus that the first array element may be undefined or 0.

SECOND VERSION OF THE PROGRAM

Some computer systems set all variables to 0 when any program is started, and others simply leave whatever value exists in that memory location from the previous program alone. In any case, the value of I is unlikely to be 1 as desired. This is a case of improper *initialization* of variables. So we add a line to the program here to set I to 1:

```
(*READ IN DATA UNTIL END OF FILE*)
I:=1;    (*<<<<<<<*)
REPEAT
    READLN(F1,X[I]);
    I:=I+1;
UNTIL EOF(F1);
```

Then we reexecute the program.

Upon running the new version BUGGY2, we find that we get the same error message

? ARRAY INDEX OUT OF BOUNDS

and we realize that we will only get somewhere with this problem if we start printing out the intermediate results in the program. So we add another line to the program to print out each value as it is read in, and to print out the value for the variable N. Thus we modify lines 1800–2300 to read:

```
REPEAT
     READLN(F1,X[I]);
     WRITELN('I=',I:2,'  X[I]=',X[I]:5:0);   (*<<<<*)
     I:=I+1;
UNTIL EOF(F1);
(*REMEMBER NUMBER READ IN*)
N:=I;
WRITELN('N=',N);     (*<<<<<*)
```

RESULTS OF THE THIRD VERSION

This time we actually get some results from our program The output of the program looks like this:

```
I= 1  X[I]=    1.
I= 2  X[I]=    2.
I= 3  X[I]=    3.
I= 4  X[I]=    4.
I= 5  X[I]=    5.
I= 6  X[I]=    6.
I= 7  X[I]=    7.
I= 8  X[I]=    8.
I= 9  X[I]=    9.
I=10  X[I]=   10.
N=   11

? ARRAY INDEX OUT OF BOUNDS

EXIT
```

This time we see that the data has been read in correctly, but that the number of data points in the read in array N has been set 1 too high since I is incremented *after* each element is read in. Here we recognize that the correct way to read from any file is using the WHILE NOT EOF (file) DO statement, so that we not only cannot read beyond the end of file, but can handle empty files without error.We then change the array index I to start at 0 and be incremented *only* if there is another array element to be read in:

```
I:=0;    (*START BELOW BEGINNING OF ARRAY*)
WHILE NOT EOF(F1) DO     (*READ IN IF THERE IS ANYTHING THERE*)
     BEGIN
     I:=I+1;       (*COUNT ACTUAL ELEMENTS READ*)
     READLN(F1,X[I]);     (*READ EACH ELEMENT*)
     END; (*WHILE*)
```

We then look further through the program for other problems.

VERSION FOUR OF THE BUGGY PROGRAM

In looking through the program for further problems, we note that the value of I is 10 at the time that N is printed out and thus that I will be 10 at line 2700 where the ODDSUM is calculated. Thus I again lacks proper *initialization.* We further note that the initial value of ODDSUM is not defined within the program and insert at line 2600 the initialization lines:

```
I:=1;
ODDSUM:=0.0;
```

and at line 3100:

```
EVENSUM:=0.0;
```

The output of the fourth version then looks like this:

```
I= 1   X[I]=    1.
I= 2   X[I]=    2.
I= 3   X[I]=    3.
I= 4   X[I]=    4.
I= 5   X[I]=    5.
I= 6   X[I]=    6.
I= 7   X[I]=    7.
I= 8   X[I]=    8.
I= 9   X[I]=    9.
I=10   X[I]=   10.
N=   10

? ARRAY INDEX OUT OF BOUNDS
```

Version four of this program still does not correct all the array indexing errors, and we need to look at the two REPEAT loops where the ODDSUM and the EVENSUM are calculated. Looking at lines 2600–2900 we see:

```
I:=1;
ODDSUM:=0;
REPEAT
   ODDSUM:=ODDSUM+X[I];
   I:=I+2;
UNTIL I=N;
```

We realize that I starts at 1, then moves upward by 2 at a time, taking on the values 3, 5, 7, 9, 11, . . ., and *never* becomes exactly equal to 10. This is a classic case of using an "equals" test when a "> =" or "< =" should have been used. We thus modify lines 2900 and 3400 to read:

```
UNTIL I>=N;
```

and remove the printout of read in values. We also note that I is not reinitialized for the EVENSUM loop and insert I:=2 before that loop at line 3100. By now, our revised program looks like this:

```
PROGRAM BUGGY5;
(*THIS PROGRAM CALCULATES THE GRONCH CONSTANT
WHICH IS EQUAL TO THE SUM OF THE ODD ELEMENTS
DIVIDED BY TWICE THE SUM OF THE EVEN ELEMENTS OF ANY ARRAY*)

CONST
        MAX=10; (*ARRAY MAXIMUM*)

VAR
        X: ARRAY[1..MAX] OF REAL;    (*ARRAY TO GRONCH ON*)
        N,I:INTEGER;     (*MAX IN ARRAY AND INDEX*)
        ODDSUM,EVENSUM,GRONCH:REAL;
        F1:TEXT;     (*INPUT TEXT FILE*)

BEGIN
        RESET(F1,'GRONCH.DAT');
(*READ IN DATA UNTIL END OF FILE*)
I:=0;    (*START I AT BEGINNING OF ARRAY*)
WHILE NOT EOF(F1) DO
    BEGIN
    I:=I+1;
        READLN(F1,X[I]);
    END;
(*REMEMBER NUMBER READ IN*)
N:=I;
WRITELN('N=',I:2);

(*DO ODDSUM*)
I:=1;    (*SET INDEX TO BOTTOM OF ARRAY*)
ODDSUM:=0;   (*AND ZERO SUM*)
REPEAT
        ODDSUM:=ODDSUM+X[I];
        I:=I+2;
UNTIL I>=N;
```

```
(*NOW DO EVEN SUM*)
I:=2;         (*RESET I TO 2 FOR THIS LOOP *)
EVENSUM:=0;
REPEAT
         EVENSUM:=EVENSUM+X[I];
         I:=I+2;
UNTIL I>=N;

(*NOW CALCULATE THE GRONCH CONSTANT*)
GRONCH:=ODDSUM/EVENSUM*2;
WRITELN('GRONCH CONSTANT= ',GRONCH);
END.
```

The output of this fifth version is:

```
N=          10
GRONCH CONSTANT= 2.5000000
```

This looks great, and it seems as if we finally got the program working. We check with our hand-calculated case and find that the correct value should be 0.41667.

SIXTH VERSION OF THE PROGRAM

The value 2.5 is clearly incorrect, and in an attempt to find our error, we print out the intermediate results ODDSUM and EVENSUM, adding an additional WRITE statement just before the final statement which prints out the value of GRONCH. Running this version, we find the values:

```
N=          10
ODDSUM= 2.5000000E+01   EVENSUM=  2.0000000E+01
GRONCH CONSTANT= 2.5000000
```

Clearly the ODDSUM is being calculated correctly, the EVENSUM is not being calculated correctly, and the GRONCH constant is also incorrect. There are, then, at least two problems to fix.

Looking at the calculation of the EVENSUM (lines 3100–3400), we see that I and EVENSUM are correctly initialized to 1 and 0, respectively, and that our loop at first "looks" all right. However, after some thought we might notice that repeating until I> =10 will exit from the REPEAT loop when I = 10, and after only X[8] had been added in, which is one element *too soon.*

We thus realize that in our haste to fix the REPEAT loop for ODDSUM, we have introduced a new bug into the loop for EVENSUM. This can be fixed by changing the REPEAT loop terminator to:

```
UNTIL I>N;
```

Then, looking at our statement where GRONCH is actually calculated, we see:

```
GRONCH:= ODDSUM/EVENSUM*2;
```

Here is the classic case of an error in the order of evaluation of arithmetic operands. In Pascal it may not be clear to the reader what is meant by that statement, and in fact we recall that Pascal evaluates such statements from left to right, first dividing and then multiplying, as if we had written

```
GRONCH:= (ODDSUM/EVENSUM)*2;
```

when we actually mean

```
GRONCH:= ODDSUM/(EVENSUM*2);
```

By changing the program to give this last statement, we are ready to execute the final version. The output of this program is:

```
ODDSUM= 2.5000000E+01  EVENSUM=  3.0000000E+01
GRONCH CONSTANT= 4.1666666E-01
```

```
PROGRAM BUGGY7;
(*THIS PROGRAM CALCULATES THE GRONCH CONSTANT
WHICH IS EQUAL TO THE SUM OF THE ODD ELEMENTS
DIVIDED BY TWICE THE SUM OF THE EVEN ELEMENTS OF ANY ARRAY*)

CONST
        MAX=10; (*ARRAY MAXIMUM*)

VAR
        X: ARRAY[1..MAX] OF REAL;    (*ARRAY TO GRONCH ON*)
        N,I:INTEGER;     (*MAX IN ARRAY AND INDEX*)
        ODDSUM,EVENSUM,GRONCH:REAL;
        F1:TEXT;     (*INPUT TEXT FILE*)

BEGIN
        RESET(F1,'GRONCH.DAT');
(*READ IN DATA UNTIL END OF FILE*)
I:=0;    (*START I AT BEGINNING OF ARRAY*)
WHILE NOT EOF(F1) DO
    BEGIN
    I:=I+1;
        READLN(F1,X[I]);
    END;
```

```
(*REMEMBER NUMBER READ IN*)
N:=I;
WRITELN('N=',I:2);

(*DO ODDSUM*)
I:=1;    (*SET INDEX TO BOTTOM OF ARRAY*)
ODDSUM:=0;  (*AND ZERO SUM*)
REPEAT
         ODDSUM:=ODDSUM+X[I];
         I:=I+2;
UNTIL I>=N;

(*NOW DO EVEN SUM*)
I:=2;          (*RESET I TO 2 FOR THIS LOOP *)
EVENSUM:=0;
REPEAT
         EVENSUM:=EVENSUM+X[I];
         I:=I+2;
UNTIL I>N;

(*NOW CALCULATE THE GRONCH CONSTANT*)
WRITELN('ODDSUM=',ODDSUM,'  EVENSUM= ',EVENSUM);
GRONCH:=ODDSUM/(EVENSUM*2);
WRITELN('GRONCH CONSTANT= ',GRONCH);
END.
```

SUMMARY

In this chapter we have seen a general debugging strategy of printing out each step of a program to see where the error is. This can be expanded to the form of printing out simple messages throughout the program, to figure out how far a complex program "gets" before it goes awry.

We have also seen the importance of knowing the initial value of variables and of reinitializing them for new loops throughout the program.

Finally we have noted that careless use of parentheses and general punctuation accounts for a number of remaining Pascal errors. A trivial, but annoying additional problem occurs in the case of an unclosed comment, which prevents some statements from being compiled at all. For example, the fragment:

```
READLN(J);   (*GET J FROM TERMINAL*
WRITELN('PRINTING OUT VALUES'); (*PRINT HEADER*)
WRITELN('J=',J);     (*LIST J*)
```

will never print out the header message "PRINTING OUT VALUES" because the comment in the first line is not terminated with a *) pair. This is not an error the compiler can find, because it considers all of the text a comment from the occurrence of the (* symbol to the next occurrence of the *) symbol which, in this case, is in the comment at the end of the next line. The compiler will sometimes indicate that there is an error on the line after the comment is finally closed with a subsequent "*)."

PROBLEM

1 The following program is supposed to read in a value for J and print it out again. No matter what value is entered for J, the same value is always printed out. Why?

```
PROGRAM ECHO;
VAR J:INTEGER;

(********************************)
PROCEDURE GETJ(I:INTEGER);
BEGIN
    READLN(I);   (*READ IN I FROM TERMINAL*)
END; (*GETJ*)
(********************************)
BEGIN   (*MAIN*)
    WRITE('ENTER J: ');
    GETJ(J);     (*READ IN J*)
    WRITELN('J=',J);
END.
```

CHAPTER 16

Using Pascal to Create New Features

One of the greatest powers of Pascal is its ability to define new features and data types in terms of the existing types. In this chapter we illustrate two such features: complex arithmetic and a "forgiving" real input routine.

COMPLEX NUMBERS

A major omission in Pascal is the complex data type for the handling of complex numbers. This was intentional, since it is perfectly possible to write procedures to carry out complex arithmetic when needed, and was done to avoid cluttering the language (and its run-time system) with seldom used features that would be "nice to have."

In order to define complex numbers, we recall that they have the general form:

$$a + bi$$

where i is the imaginary number representing the square root of -1.

Complex numbers, then, are really two real values, one of which is assumed to be the coefficient of 1 and the other the coefficient of i. We can define a new data type as a record of two real numbers by:

```
TYPE
   COMPLEX=
      RECORD
      R,I: REAL;
      END: (*RECORD*)
```

and then manipulate whole records as usual. When we wish to refer to either coefficient individually, it will be as A.R or A.I if A is of type COMPLEX.

COMPLEX ARITHMETIC

Complex addition, then, is carried out through the procedure CPADD, in which we recognize the complex sum:

$$(a + bi) + (c + di) = (a + c) + (b + d)i$$

The procedure is written placing the result in a new record C so that the original values are not disturbed.

In a similar fashion we know that the product of two complex numbers is given by:

$$(a + bi)(c + di) = (ac - bd) + (bc + ad)i$$

and this equation is implemented in the procedure CPMULT.

Complex division is carried out by recognizing that the numerator and the denominator must be multiplied by the complex conjugate of the denominator $(c - di)$ so that there will be only a real part in the denominator. The result will be given by:

$$\frac{(a + bi)}{(c + di)} = \frac{(a + bi)(c - di)}{(c + di)(c - di)} = \frac{(ac + bd) + (bc - ad)i}{(c^2 + d^2)}$$

This latter equation can be treated in terms of a call to the complex multiplication routine to calculate the numerator. This is shown in the program below, which reads in two complex numbers and prints out their sum, product, and quotient. Note the use of the formatting constants F and D to simplify the WRITELN expressions.

```
PROGRAM COMPLEXER;
(*THIS PROGRAM CALCULATES COMPLEX SUMS,PRODUCTS
AND QUOTIENTS AND PRINTS THEM OUT FOR TWO ENTERED COMPLEX NUMBERS*)

CONST
    F=12;    D=4;      (*FIELDWIDTH AND DEC PLACES*)
TYPE
    COMPLEX=
        RECORD
        R,I:REAL;      (*TWO REAL NUMBERS MAKE UP COMPLEX NUMBER*)
        END;

VAR X,Y,Z:COMPLEX;
(*********************************************)
PROCEDURE CPMULT(A,B:COMPLEX; VAR C:COMPLEX);
(*PERFORMS COMPLEX MULTIPLICATIONS
     C:=A*B;
*)
BEGIN
     C.R:=A.R*B.R -A.I*B.I; (*CALC  REAL PART*)
     C.I:= A.I*B.R + A.R*B.I; (*CALC IMAG PART*)
END;      (*CPMULT*)
(*********************************************)

PROCEDURE CPADD(A,B:COMPLEX; VAR C:COMPLEX);
(*PERFORMS COMPLEX ADDITION OF TYPE
     C:=A+B;
*)
BEGIN
     C.R:=A.R+B.R;
     C.I:=A.I+B.I;
END;      (*CPADD*)
```

```
(*******************************************)
PROCEDURE CPDIV(A,B:COMPLEX; VAR C:COMPLEX);
(*PERFORMS COMPLEX DIVISION OF TYPE
    C:=A/B;
*)
VAR DENOM:REAL;

BEGIN
    DENOM:=SQR(B.R)+SQR(B.I);
    B.I:=-B.I;  (*COMPLEX CONJ, DOES NOT CHANGE B IN CALLING PGM*)
    CPMULT(A,B,B);
    IF DENOM=0.0 THEN WRITELN('DIVISION BY 0! ')
    ELSE
        BEGIN
        C.R:=B.R/DENOM;
        C.I:=B.I/DENOM;
        END;(*ELSE*)
END;    (*CPDIV*)

(*******************************************)
BEGIN   (*MAIN*)
    WRITE('ENTER X.REAL<SPACE>X.IMAG<RETURN>: ');
    READLN(X.R,X.I);
    WRITE('ENTER Y.REAL & Y.IMAG: ');
    READLN(Y.R,Y.I);
(*COMPLEX ARITHMETIC*)
    CPADD(X,Y,Z);    (*COMPLEX ADDITION*)
    WRITELN('ADDN: ',Z.R:F:D,'  +',Z.I:F:D,'I');
    CPMULT(X,Y,Z);   (*COMPLEX MULTIPLICATION*)
    WRITELN('MULT: ',Z.R:F:D,'  +',Z.I:F:D,'I');
    CPDIV(X,Y,Z);    (*COMPLEX DIVISION*)
    WRITELN('DIVN: ',Z.R:F:D,'  +',Z.I:F:D,'I');
END.
```

For many more exotic methods using complex numbers, see Ref. (1) which illustrates Bessel and Hankel functions.

A "FORGIVING" REAL INPUT ROUTINE

It is perfectly possible in Pascal to have a real input routine that does not require that the user type a digit on both sides of the decimal point as standard READ routines require. The version shown here allows almost any readable real format and reads up through the first illegal character without issuing an error message.

In this program various input formats are allowed, such as:

12
12.
.12
12.E3
12.0E3

12E3
.12E3

and so forth.

This is accomplished by checking each entered character for membership in the set of digits from '0' to '9' and repeatedly getting digits until a character is not a member of that set. Then checks are made for '.', 'E', or an illegal character.

The program starts in the procedure READREAL by skipping any leading spaces as usual for Pascal input routines. Then it allows an optional sign if desired. Following the sign can come digits until a '.', 'E', or illegal character is entered, or a '.' can be entered directly. If a '.' is entered, the routine accepts more digits until an illegal character or an 'E' is entered. If the character 'E' is entered, the EXPONENT routine allows entry of an optional sign and accepts digits until an illegal character is entered.

As soon as an illegal character is found, the routine exists to the calling program, with the calculated real value in NUM. The function DIG is simply used to convert characters to numerical digits by subtracting the value of the character '0' from each entered character which is a member of the set of digits.

This routine also illustrates passing a text file variable to procedures. This allows the READREAL and EXPONENT routines to read from any text file, including the terminal or INPUT file. Note that the file identifier F is an argument to READREAL and EXPONENT, and that in the main program the calling parameter is INPUT, or the terminal in DEC-10 and UCSD Pascal. Note also that the file variable must be a reference parameter (using VAR) and not a value parameter.

```
PROGRAM READR;
(*ILLUSTRATES THE USE OF THE FORGIVING INPUT ROUTINE READREAL*)
VAR DIGITS: SET OF CHAR;     (*SET TO TEST FOR ACTUAL DIGIT*)
R:REAL; (*INPUT NUMBER*)

(**********************************************)
FUNCTION DIG(A:CHAR):INTEGER;
(*RETURNS INTEGER EQUAL TO CHARACTER A*)
BEGIN
    DIG:=ORD(A)-ORD('0');
END;
(**********************************************)
FUNCTION FSIGN(VAR F:TEXT; VAR C:CHAR):BOOLEAN;
(*RETURNS TRUE IF MINUS SIGN NEXT
AND TRUE IF PLUS SIGN OR NO SIGN*)
BEGIN
    IF C='-' THEN FSIGN:=TRUE
        ELSE FSIGN:=FALSE;
```

```
    IF (C='+') OR (C='-') THEN READ(F,C);
END;     (*FSIGN*)
(**********************************************)
PROCEDURE EXPONENT(VAR F:TEXT; VAR N:REAL);
(*GETS THE EXPONENT PART, IF ANY*)

VAR
    EXNUM:INTEGER;   (*EXPONENT VALUE*)
    C:CHAR;      (*CHARACTER READ IN*)
    EXSIGN:BOOLEAN; (*SIGN OF EXPONENT*)

BEGIN
READ(F,C);
EXNUM:=0;
(*FIRST CHECK FOR SIGN*)
EXSIGN:=FSIGN(F,C);

WHILE (C IN DIGITS) AND NOT EOLN(F) DO
    BEGIN
    EXNUM:=EXNUM*10 +DIG(C);
    READ(F,C);
    END;(*WHILE*)
(*USE LAST CHARACTER IF NOT ILLEGAL*)
IF C IN DIGITS THEN
    EXNUM:=EXNUM*10 + DIG(C);

(*THEN SET EXPONENT SIGN*)
IF EXSIGN THEN EXNUM:=-EXNUM;

(*MULTIPLY BY THAT POWER OF TEN*)
N:=N*EXP(EXNUM*LN(10));
END;
(**********************************************)
PROCEDURE READREAL(VAR F:TEXT; VAR NUM:REAL);
(*THIS PROCEDURE READS A REAL NUMBER FROM
THE TERMINAL OR ANY TEXT FILE IN A MOST FORGIVING FASHION.
NO DECIMAL POINT IS NEEDED FOR INTEGRAL VALUES,
NO DIGITS TO THE RIGHT OF THE DECIMAL ARE
NEEDED AND A + SIGN IS OPTIONAL, BUT ALLOWED.*)

VAR
    SIGN:BOOLEAN;
    C:CHAR;
    DECPLACES:REAL;


BEGIN
DIGITS:=['0'..'9']; (*INITIALIZE SET*)
NUM:=0.0;    (*START WITH ZERO VALUE*)
(*SKIP LEADING SPACES*)
REPEAT
    READ(F,C);
UNTIL C<>' ';

(*NOW SET SIGN*)
SIGN:=FSIGN(F,C);

(*NOW GET DIGITS UP TO DECIMAL POINT*)
WHILE (C IN DIGITS) AND NOT EOLN(F) DO
    BEGIN
    NUM:=NUM*10 +DIG(C);
    READ(F,C);
    END;(*WHILE*)
```

```
(*GET LAST CHARACTER*)
IF C IN DIGITS THEN
     NUM:=NUM*10 + DIG(C);

(*DIGITS CAN BE TERMINATED WITH 'E' OR '.' OR ILLEGAL CHAR
CHECK BOTH POSSIBILITIES -ILLEGAL CHAR FALLS THROUGH*)

IF (C='.') OR (C='E') THEN  (*RESTRICT CASE*)
CASE C OF
'.': BEGIN
     DECPLACES:=10.0;
     READ(F,C);
     WHILE (C IN DIGITS) AND NOT(EOLN(F)) DO
          BEGIN
          NUM:=NUM + DIG(C)/DECPLACES;
          DECPLACES:=DECPLACES*10.0;
          READ(F,C);
          END;(*WHILE*)
     IF C IN DIGITS THEN      (*USE LAST CHAR*)
          NUM:=NUM +DIG(C)/DECPLACES;
     IF C='E' THEN EXPONENT(F,NUM);
     END;(*'.'*)

'E':     EXPONENT(F,NUM);
END;(*CASE*)

(*SETS SIGN AND EXIT*)
IF SIGN THEN NUM:=-NUM;
END;(*READREAL*)
(*************************************************)
(*MAIN PROGRAM*)
BEGIN
REPEAT
     READREAL(INPUT,R);   (*READ A REAL NUMBER*)
     WRITELN('R= ',R);
UNTIL R=0.0;
END.
```

PROBLEMS

1 The procedure READREAL given above differs from the standard routines in one aspect: the routine exits *after* reading the terminating character. How could you rewrite READREAL to exit when the *next* character is a terminating character?

REFERENCES

1 *Pascal News,* 17, 47–51, March 1980.

CHAPTER 17

Matrix Manipulation in Pascal

A *matrix* is a square or rectangular array of numbers which has certain specific mathematical uses and properties. Matrices can be added, subtracted, multiplied, inverted, and diagonalized using standard rules of matrix algebra. The matrix inversion and diagonalization methods have specific uses in the physical sciences and are thus covered in detail in this chapter. In order to appreciate these methods, however, we will start by giving a few fundamental properties of matrices.

PROPERTIES OF MATRICES

If we consider the general matrix

$$\begin{bmatrix} a_{11} & a_{12} & a_{13} \\ a_{21} & a_{22} & a_{23} \\ a_{31} & a_{32} & a_{33} \end{bmatrix}$$

we see that we have written a square array of numbers having subscripts referring to their rows and columns, in that order. Matrices can be represented by a single capital letter, such as A. If A and B are two 2×2 matrices:

$$A = \begin{bmatrix} a_{11} & a_{12} \\ a_{21} & a_{22} \end{bmatrix} \quad B = \begin{bmatrix} b_{11} & b_{12} \\ b_{21} & b_{22} \end{bmatrix}$$

then the matrix sum $A + B$ is given by adding together the corresponding elements of each matrix, and is defined only for matrices having the same number of rows and columns:

$$A + B = \begin{bmatrix} a_{11} + b_{11} & a_{12} + b_{12} \\ a_{21} + b_{21} & a_{22} + b_{22} \end{bmatrix} \tag{17.1}$$

Similar rules apply for subtraction.

Multiplication of matrices is only defined if the number of columns in the first matrix is equal to the number of rows in the second matrix. Then the product $A \times B$ is given by multiplying each element in each row of the first matrix by the corresponding element in that column of the second matrix and summing these products for a given row and column. This can be written as:

$$C = A \times B$$

or if A is a matrix of m rows and p columns and B is a matrix of p rows and n columns:

$$C_{ij} = \sum_{k=1}^{p} a_{ik} b_{kj} \tag{17.2}$$

For the matrices

$$A = \begin{bmatrix} 1 & 2 \\ 3 & 4 \end{bmatrix} \quad B = \begin{bmatrix} 5 & 6 \\ 7 & 8 \end{bmatrix}$$

the product $A \times B$ will be given as a 2×2 matrix consisting of:

$$\begin{bmatrix} (1\times5) + (2\times7) & (1\times6) + (2\times8) \\ (3\times5) + (4\times7) & (3\times6) + (4\times8) \end{bmatrix} = \begin{bmatrix} 19 & 22 \\ 43 & 50 \end{bmatrix}$$

Multiplication is always row by column—with the rows of the first matrix multiplied by the columns of the second to produce the new matrix element. Note that matrix multiplication is not commutative:

$$A \times B \neq B \times A$$

Matrices cannot be divided as such, but they can be inverted. To understand the inverse, we first define the unit matrix I as being a matrix whose diagonal elements are 1's and all off-diagonal elements are 0's. By diagonal we always mean that diagonal that runs from top left to bottom right. A 4×4 unit matrix would be written as:

$$\begin{bmatrix} 1 & 0 & 0 & 0 \\ 0 & 1 & 0 & 0 \\ 0 & 0 & 1 & 0 \\ 0 & 0 & 0 & 1 \end{bmatrix}$$

The *inverse* of a matrix A is defined as that matrix A^{-1} such that

$$A \times A^{-1} = I, \text{ the unit matrix} \tag{17.3}$$

Matrix inversion is a well-understood process which can be used to solve a series of simultaneous equations, as we will see below.

SOLVING SIMULTANEOUS EQUATIONS BY MATRIX INVERSION

Let us consider a series of simultaneous equations such as:

$$\begin{aligned} a_1x_1 + b_1x_2 + c_1x_3 &= k_1 \\ a_2x_1 + b_2x_2 + c_2x_3 &= k_2 \\ a_3x_1 + b_3x_2 + c_3x_3 &= k_3 \end{aligned} \tag{17.4}$$

This can be written in matrix form as

$$\begin{bmatrix} a_1 & b_1 & c_1 \\ a_2 & b_2 & c_2 \\ a_3 & b_3 & c_3 \end{bmatrix} \begin{bmatrix} x_1 \\ x_2 \\ x_3 \end{bmatrix} = \begin{bmatrix} k_1 \\ k_2 \\ k_3 \end{bmatrix} \tag{17.5}$$

or just as

$$Mx = k \tag{17.6}$$

where M represents the matrix of coefficients, and x and k represent the column matrices of variables and constants, respectively.

Such a series of simultaneous equations can be solved by creating a matrix of the coefficients and an array of the constants, inverting the matrix M and multiplying that inverted matrix by the constant column matrix k. The resulting column matrix will be the values of the unknown x's. In other words,

$$(M^{-1})\ k = x \tag{17.7}$$

which is what we might expect from multiplying Eq. 17.7 by M^{-1}.

Inversion of a matrix is a fairly complicated process in which rows and columns are swapped, and the largest element in a given row is found and used as a *pivot element* to divide through by. In the program below, a matrix is read in from the MATINP file starting with a line containing the matrix dimension, and then the coefficient matrix and constant matrix are read in row by row. The matrix is then inverted and multiplied by the constant matrix, and the final values of the x's are printed out.

For this method to work it is *essential* that none of the diagonal elements be zero, or division by zero will occur. This can be avoided, if zero coefficients occur, by simply rearranging the rows of the input matrix.

```
PROGRAM SIMULSOLVE;
(*THIS PROGRAM READS IN A SET OF VALUES
FROM THE INPUT FILE HAVING THE FORMAT:
     n
     a11 a12 a13...a1n  k1
     a21 a22 a23...a2n  k2
     :              :   :
     :
     an1 an2 an3...ann  kn
IT INVERTS THE A MATRIX, AND MULTIPLIES IT BY THE K COLUMN MATRIX
TO PRODUCE THE SOLUTIONS FOR THE SIMULTANEOUS EQUATIONS.
IF THE MATRIX IS SINGULAR, THE DETERMINANT
IS ZERO AND AN ERROR MESSAGE IS PRINTED. *)

(* LABEL 99; *) (*ILLUSTRATES HOW GOTO COULD ALSO BE USED*)
CONST MATMAX=10;
F=10;   W=3;

TYPE
     MATDIM=1..MATMAX;
     XARY=ARRAY[MATDIM,MATDIM] OF REAL;

VAR
     I,J,SIZE:MATDIM;
     X:XARY;
     D:REAL; (*DETERMINANT RETURNED HERE*)
     SOLNS,SUMS:ARRAY[MATDIM]OF REAL;
     F1:TEXT;          (*MATINP INPUT FILE*)

(**************************************************)
PROCEDURE MATINV(VAR A:XARY; N:MATDIM; VAR DETERM:REAL);
(* MATRIX INVERSION PROCEDURE:
     A IS A N X N MATRIX
     N IS THE SIZE OF THE MATRIX
     DETERM IS THE DETERMINANT OF THE MATRIX
THIS PROCEDURE INVERTS A REAL SYMMETRIC MATRIX OF
DIMENSION MATDIM x MATDIM, FOLLOWING THE PROCEDURE
GIVEN BY JOHNSON (1)   *)
(*PROCEDURE ASSUMES NON-ZERO DIAGONAL ELEMENTS!!
IF THE MATRIX IS SINGULAR, DETERM IS 0
AND THE PROCEDURE IS TERMINATED*)

CONST
     TOL=1.0E-34;     (*VERY CLOSE TO ZERO*)

VAR
     NSWAP:INTEGER;
     IPV:ARRAY[MATDIM,1..3] OF INTEGER;
     I,J,K,L,ICOLUMN,JCOLUMN,IROW,JROW,L1:MATDIM;
     T,PIVOT,SWAP,AMAX:REAL;

BEGIN
DETERM:=1.0;     (*START DETERMINANT AS 1.0*)
FOR J:=1 TO N DO IPV[J,3]:=0;    (*INITIALIZE*)
I:=1;    (*DO FROM I TO N*)
REPEAT   (*UNTIL I>N OR DETERMINANT BECOMES 0 *)
(*SEARCH FOR PIVOT ELEMENT*)
     AMAX:=0.0;
     FOR J:=1 TO N DO
          BEGIN
```

```
    IF IPV[J,3]<>1.0 THEN
        BEGIN
        FOR K:=1 TO N DO
            BEGIN
            IF IPV[K,3]<>1.0 THEN
                IF AMAX < ABS(A[J,K]) THEN
                    BEGIN
                    IROW:=J;      (*SAVE ROW AND COLUMN*)
                    ICOLUMN:=K;
                    AMAX:=ABS(A[J,K]);
                    END;     (*IF AMAX*)
            END; (*FOR K*)
        END; (*IF IPV[J,3]*)
    END; (*FOR J*)

(*IF PIVOT ELEMENT IS NEAR 0 - DETERMINANT WILL BE 0*)
IF AMAX < TOL THEN DETERM := 0.0
(*IF DETERM=0 THEN GOTO 99      >>>HERE WE COULD USE A GOTO STATEMENT!!!*)
ELSE
    BEGIN   (*DO THIS ONLY IF MATRIX HAS NOT BECOME SINGULAR*)
    IPV[ICOLUMN,3]:=IPV[ICOLUMN,3]+1;
    IPV[I,1]:=IROW;
    IPV[I,2]:=ICOLUMN;
    (*INTERCHANGE ROWS TO PUT PIVOT ELEMENT ON DIAGONAL*)
    IF IROW<>ICOLUMN THEN
        FOR L:=1 TO N DO
            BEGIN
            SWAP:=A[IROW,L];
            A[IROW,L]:=A[ICOLUMN,L];
            A[ICOLUMN,L]:=SWAP;
            END;(*FOR L*)

    (*DIVIDE PIVOT ROW BY PIVOT ELEMENT*)
    PIVOT:=A[ICOLUMN,ICOLUMN];
    DETERM:= DETERM * PIVOT;
    A[ICOLUMN,ICOLUMN]:=1.0;
        FOR L:=1 TO N DO
            A[ICOLUMN,L]:=A[ICOLUMN,L]/PIVOT;

    (*REDUCE THE NON-PIVOT ROWS BY SUBTRACTION*)
    FOR L1:=1 TO N DO
        IF L1<>ICOLUMN THEN      (*AVOID DIAGONAL*)
            BEGIN
            T:=A[L1,ICOLUMN];
            A[L1,ICOLUMN]:=0.0;
            FOR L:=1 TO N DO
                A[L1,L]:=A[L1,L]-A[ICOLUMN,L]*T;
            END;(*IF*)
    I:=I+1; (*GO ON TO NEXT I*)
    END; (*ELSE IF AMAX<TOL*)
UNTIL (I>N) OR (DETERM = 0.0);

    (*INTERCHANGE THE COLUMNS AND MODIFY DETERMINANT *)
    NSWAP:=0;    (* THIS IS SIGN FLAG*)
    IF DETERM <> 0 THEN BEGIN   (*DO THIS ONLY IF MATRIX NOT SINGULAR*)
    FOR I:=1 TO N DO
        BEGIN
        L:=N-I+1;
        IF IPV[L,1]<>IPV[L,2] THEN
            BEGIN
            JROW:=IPV[L,1];
            JCOLUMN:=IPV[L,2];
            NSWAP:=NSWAP+1; (*COUNT SWAPS*)
            FOR K:=1 TO N DO
```

```
                BEGIN
                SWAP:=A[K,JROW];
                A[K,JROW]:=A[K,JCOLUMN];
                A[K,JCOLUMN]:=SWAP;
                END; (*FOR K*)
            END(*IF IPV*)
        END;(*FOR I*)
IF ODD(NSWAP) THEN SWAP:=-1
        ELSE SWAP:=1;
DETERM:=DETERM*SWAP;
END;      (*IF DETERM<>0 *)
(* 99: *)         (*THE FOLLOWING WOULD BE THE LABELED STATEMENT*)
END;      (*MATINV*)
(**************************************************)
BEGIN (*MAIN PROGRAM*)
RESET(F1,'MATINP    ');
(*GET MATRIX SIZE TO ENTER*)
READLN(F1,SIZE);     (*READ SIZE FROM INPUT FILE*)

(*NOW ALLOW ENTRY OF ELEMENTS BY ROWS*)
FOR I:=1 TO SIZE DO
     BEGIN
     FOR J:=1 TO SIZE DO
          BEGIN
          READ(F1,X[I,J]);
          END;
     READ(F1,SUMS[I]);
     END;(*FOR I*)
     MATINV(X,SIZE,D);    (*INVERT MATRIX*)
(*PRINT ERROR MESSAGE IF MATRIX WAS SINGULAR*)
IF D=0.0 THEN WRITELN('MATRIX IS SINGULAR')

(*OTHERWISE PRINT OUT THE RESULT*)

ELSE
BEGIN
     FOR I:=1 TO SIZE DO
     BEGIN
     FOR J:=1 TO SIZE DO
          WRITE(X[I,J]:10:5);
          WRITELN;
          END;(*FOR I*)
     (*DO MATRIX PRODUCT OF SUMS WITH INVERSE
     TO GET THE SOLUTIONS OF X's*)
     FOR I:=1 TO SIZE DO
          BEGIN
          SOLNS[I]:=0.0;   (*INITIALIZE PRODUCTS TO 0*)
          FOR J:=1 TO SIZE DO
               BEGIN
               SOLNS[I]:=SOLNS[I]+X[I,J]*SUMS[J];
               END;(*FOR J*)
          END;(*FOR I*)
     (*WRITE OUT PRODUCTS*)
     WRITELN('SOLUTIONS OF EQUATIONS');
     FOR I:=1 TO SIZE DO
          WRITELN(I:2,SOLNS[I]:F:W);
     END;     (*ELSE*)
     END.
```

A *singular* matrix has a zero determinant. Singular matrices usually occur when the simultaneous equations are not independent, such as:

$$x + y + z = 5$$
$$2x + 2y + 2z = 10$$

In this case, of course, there is insufficient information for a solution, and the inversion program quits with DETERM set to zero.

For example, suppose we have the equations:

$$x + 2y + 3z = 26$$
$$5x + 3y + z = 32$$
$$2x + y + 2z = 20$$

We simply make up an input file that looks like this:

```
3
1.0  2.0  3.0  26.0
5.0  3.0  1.0  32.0
2.0  1.0  2.0  20.0
```

and run the program above to get the inverted matrix and the solutions for x. In this case we will find that

$$x = 3$$
$$y = 4$$
$$z = 5$$

MATRIX DIAGONALIZATION

In scientific problems a system of linear equations often occurs which can be solved by matrix diagonalization. The following discussion follows that given by Dickson (2). Such a series of equations might have the form

$$Ax = \lambda x \tag{17.8}$$

where λ is a constant, A is a matrix of coefficients, and x is a column vector of unknowns. Constant values λ which satisfy this equation are termed *eigen-*

values, and the corresponding x's are called *eigenvectors*. We can rewrite the above as

$$Ax - \lambda x = 0$$

or

$$(A - \lambda I)x = 0$$

or

$$Ax = \lambda Ix \tag{17.9}$$

Writing Eq. 17.9 as simple 3 × 3 matrices we can write:

$$\begin{bmatrix} a_{11}-\lambda & a_{12} & a_{13} \\ a_{21} & a_{22}-\lambda & a_{23} \\ a_{31} & a_{32} & a_{33}-\lambda \end{bmatrix} \begin{bmatrix} x_1 \\ x_2 \\ x_3 \end{bmatrix} = 0 \tag{17.10}$$

or as:

$$\begin{bmatrix} a_{11} & a_{12} & a_{13} \\ a_{21} & a_{22} & a_{23} \\ a_{31} & a_{32} & a_{33} \end{bmatrix} \begin{bmatrix} x_1 \\ x_2 \\ x_3 \end{bmatrix} = \begin{bmatrix} \lambda & 0 & 0 \\ 0 & \lambda & 0 \\ 0 & 0 & \lambda \end{bmatrix} \begin{bmatrix} x_1 \\ x_2 \\ x_3 \end{bmatrix} \tag{17.11}$$

There usually are several eigenvalues which are solutions to such a system and there may be as many as n solutions for an $n \times n$ system. Such systems are usually solved by determinants when done be hand.

A *determinant* is a square array of numbers which can be evaluated by a certain set of rules. By contrast, a matrix is a rectangular array which does not have a single value that can be found by some sort of mathematical process. In the two matrix equations above, we see both types of arrays. Eq. 17.10 shows a determinant multiplied by x's equal to 0, and Eq. 17.11 shows a matrix of a's multiplied by a column matrix of x's equal to λ times the unit matrix multiplied by the x column matrix.

We find that by computer, systems of linear equations can be solved most easily by *diagonalizing* the matrix, or by subtracting values from rows and columns in an orderly fashion such that the result is a matrix whose only nonzero values lie on the diagonal. These values then are found to be the eigenvalues solving the equation, and a second array which starts out as a unit matrix, transformed in the same way, will contain the eigenvectors.

The program below allows entry of a matrix from the file DIAGIN which can then be solved by diagonalization. The eigenvalues and eigenvectors are then printed out. The diagonalization process follows the method of Jacobi as described in Ralston and Wilf (3).

```
PROGRAM DIAGNL;
(*THIS PROGRAM READS IN A SHORT MATRIX FROM THE FILE DIAGIN ,
DIAGONALIZES IT, AND PRINTS OUT THE DIAGONALIZED MATRIX
AND THE CORRESPONDING EIGENVECTOR MATRIX *)

CONST MATMAX=30;     (*DIMENSION OF SQUARE MATRIX*)
    F=10;   D=3;          (*FOR OUTPUT FORMATTING*)

TYPE
     MATDIM=1..MATMAX;    (*SUBRANGE TYPE.. SUBSCRIPT VARIABES*)
     XARY=ARRAY[MATDIM,MATDIM] OF REAL;

VAR
     H,U:XARY;
     I,J,SIZE:MATDIM;
     F:TEXT; (*INPUT FILE DIAGIN*)
(*****************************************************)
PROCEDURE MDUMP(VAR X:XARY);
(*DUMPS OUT THE MATRIX ON THE TERMINAL*)
VAR I,J:MATDIM;

BEGIN
     FOR I:=1 TO SIZE DO
          BEGIN
          FOR J:=1 TO SIZE DO
          WRITE(X[I,J]:F:D);
          WRITELN;
          END;
WRITELN;
END;
(*****************************************************)
PROCEDURE MATDIAG(VAR A,S:XARY;N:MATDIM);
(*THIS PROCEDURE DIAGONALIZES A REAL SYMMETRIC
MATRIX OF DIMENSION MATDIM x MATDIM, FOLLOWING
THE METHOD OF JACOBI. THE FLOWCHART
AND VARIABLE NAMES ARE THOSE SUGGESTED BY
RALSTON AND WILF (3)  *)

VAR
     I,J,P,Q:MATDIM;
     NU,NUFINAL,RHO,LAMBDA,MU,OMEGA,SINE,COSINE,TEMP,TEMP2,
     CO2,SI2,SICO,SICO2,APP,AQQ,APQ:REAL;
     OFFDIAGFOUND:BOOLEAN;

BEGIN
RHO:=1.0E-7;     (*SMALL VALUE TO REPRESENT 0 OFF DIAGONAL*)
(*GENERATE AN IDENTITY MATRIX
HAVING OFF-DIAGONAL ELEMENTS 0'S AND
DIAGONAL ELEMENTS 1'S*)
FOR I:=1 TO N DO
```

```
    FOR J:=1 TO N DO
         IF I<>J THEN S[I,J]:=0.0
              ELSE S[I,J]:=1.0;
(*COMPUTE INITIAL NORM OF MATRIX*)
NU:-0.0;
FOR I:-1 TO N DO
    FOR J:=1 TO N DO
    IF I<>J THEN NU:=NU+SQR(A[I,J]);
NU:=SQRT(NU);
NUFINAL:=NU*RHO/N;
REPEAT  (*UNTIL NU <=NUFINAL AND NO OFF-DIAGONALS FOUND >NU*)
NU:=NU/N;   (*REDUCE NU EACH TIME UNTIL VERY SMALL*)
FOR Q:=2 TO N DO
    BEGIN
    P:=1;
    REPEAT  (*UNTIL P> Q-1 *)
         OFFDIAGFOUND:=FALSE;
         IF ABS(A[P,Q]) > NU THEN
              BEGIN
              OFFDIAGFOUND:=TRUE;
              (*SAVE ELEMENTS OF PIVOTAL SET*)
              APP:=A[P,P];     (*THESE ARE ANNTHILATED BELOW*)
              AQQ:=A[Q,Q];     (*AND THUS SAVE CONTINUAL ARRAY CALCN*)
              APQ:=A[P,Q];
              LAMBDA:=-APQ;
              MU:=(APP-AQQ)/2.0;
              IF (MU=0) OR (MU<RHO) THEN OMEGA:=-1
              ELSE
                   BEGIN
                   OMEGA:=LAMBDA/SQRT(SQR(LAMDDA) + SQR(MU));
                   IF MU<0 THEN OMEGA:=-OMEGA;
                   END; (*ELSE*9
              SINE:=OMEGA/SQRT(2*(1+SQRT(1-SQR(OMEGA))));
              COSINE:=SQRT(1-SQR(SINE));
              FOR I:=1 TO N DO
                   BEGIN
                   TEMP:=A[I,P]*COSINE-A[I,Q]*SINE;
                   A[I,Q]:=A[I,P]*SINE+A[I,Q]*COSINE;
                   A[I,P]:=TEMP;
                   TEMP:=S[I,P]*COSINE-S[I,Q]*SINE;
                   S[I,Q]:=S[I,P]*SINE+S[I,Q]*COSINE;
                   S[I,P]:=TEMP;
                   END; (*FOR I*)
                   (*NOW DO THE DIAGONAL ELEMENTS*)
                   CO2:=SQR(COSINE);    (*CALC CONSTANTS USED BELOW*)
                   SI2:-SQR(SINE);
                   SICO:=SINE*COSINE;
                   SICO2:=2*SICO*APQ;
                   A[P,P]:=APP*CO2+AQQ*SI2 -SICO2;
                   A[Q,Q]:=APP*SI2 + AQQ*CO2 + SICO2;
                   A[P,Q]:=(APP-AQQ)*SICO + APQ*(CO2-SI2);
                   A[Q,P]:=A[P,Q];
                   FOR I:=1 TO N DO
                        BEGIN
                        A[P,I]:=A[I,P];
                        A[Q,I]:-A[I,Q];
                        END; (*FOR I*)
                   END;(*IF ABS(A[P,Q])  *)
              P:=P+1; (*GO ON TO NEXT ROW*)
              UNTIL P> (Q-1);
         END;(*FOR Q*)
    UNTIL (NU<=NUFINAL) AND (NOT OFFDIAGFOUND);
    END;
```

```
(*************************************************)
BEGIN    (*MAIN*)
(*READ IN MATRIX FROM DIAGIN FILE*)
RESET(F,'DIAGIN    ');
READLN(F,SIZE); (*GET MATRIX DIMENSION*)
FOR I:=1 TO SIZE DO
     BEGIN
     FOR J:=1 TO SIZE DO
     READ(F,H[I,J]);
     READLN(F);
     END;
MATDIAG(H,U,SIZE);    (*DIAGONALIZE IT*)
MDUMP(H);(*AND DUMP IT OUT AGAIN*)
WRITELN;
MDUMP(U);
END.
```

MATRIX DIAGONALIZATION IN MOLECULAR ORBITAL THEORY

As an example of the use of a matrix diagonalization routine, let us consider the problem of simple Hückel molecular orbital theory energy level determination. The theory says that we can determine the energy levels of a particular conjugated molecule by considering only the overlap in the π-system and writing down a determinant related to the connectivity of the molecule, where all the diagonal elements are $\alpha - E$ and those off-diagonal elements that relate to atoms at actual bond-forming distances are equal to β. The remaining off-diagonal elements are zero. The values of α and β are related to the Coulomb integrals and bond energies, respectively, and are not in general evaluated but merely compared.

For the simple molecule butadiene,

$$CH_2{=}CH{-}CH{=}CH_2$$

we can write the determinant as:

$$\begin{vmatrix} \alpha - E & \beta & 0 & 0 \\ \beta & \beta - E & \beta & 0 \\ 0 & \beta & \alpha - E & \beta \\ 0 & 0 & \beta & \alpha - E \end{vmatrix} = 0 \qquad (17.12)$$

where the off-diagonal elements are β only where atoms are connected. Thus 1–2, 2–3, and 3–4 elements contain β's, but 1–4 and 1–3 do not.

We then make the simplification that

$$x = \frac{\alpha - E}{\beta} \qquad (17.13)$$

and divide Eq. 17.12 by β, forming the new determinant:

$$\begin{vmatrix} x & 1 & 0 & 0 \\ 1 & x & 1 & 0 \\ 0 & 1 & x & 1 \\ 0 & 0 & 1 & x \end{vmatrix} = 0 \qquad (17.14)$$

This determinant can be converted into a matrix of the form:

$$\begin{bmatrix} 0-\lambda & 1 & 0 & 0 \\ 1 & 0-\lambda & 1 & 0 \\ 0 & 1 & 0-\lambda & 1 \\ 0 & 0 & 1 & 0-\lambda \end{bmatrix} \begin{bmatrix} x_1 \\ x_2 \\ x_3 \\ x_4 \end{bmatrix} = 0 \qquad (17.15)$$

or as a matrix product of the form:

$$\begin{bmatrix} 0 & 1 & 0 & 0 \\ 1 & 0 & 1 & 0 \\ 0 & 1 & 0 & 1 \\ 0 & 0 & 1 & 0 \end{bmatrix} \begin{bmatrix} x_1 \\ x_2 \\ x_3 \\ x_4 \end{bmatrix} = \begin{bmatrix} \lambda & 0 & 0 & 0 \\ 0 & \lambda & 0 & 0 \\ 0 & 0 & \lambda & 0 \\ 0 & 0 & 0 & \lambda \end{bmatrix} \begin{bmatrix} x_1 \\ x_2 \\ x_3 \\ x_4 \end{bmatrix} \qquad (17.16)$$

Then, by diagonalizing the matrix on the left of Eq. 17.16, we can find the eigenvalues λ, and thus the energy levels of butadiene. Diagonalizing this matrix using the above program, we get the following diagonalized matrix whose diagonal elements are the eigenvalues:

$$\begin{bmatrix} 1.6180 & 0.0000 & 0.0000 & 0.0000 \\ 0.0000 & 0.6180 & 0.0000 & 0.0000 \\ 0.0000 & 0.0000 & -0.6180 & 0.0000 \\ 0.0000 & 0.0000 & 0.0000 & -1.6180 \end{bmatrix}$$

and the following eigenvectors:

$$\begin{bmatrix} 0.3717 & -0.6015 & 0.6015 & -0.3717 \\ 0.6015 & -0.3717 & -0.3717 & 0.6015 \\ 0.6015 & 0.3717 & -0.3717 & -0.6015 \\ 0.3717 & 0.6015 & 0.6015 & 0.3717 \end{bmatrix}$$

Briefly, the diagonal values or eigenvalues can then be substituted in Eq. 17.13 to given the energy levels in terms of α and β:

$$\alpha - 1.618\beta$$
$$\alpha - 0.618\beta$$
$$\alpha + 0.618\beta$$
$$\alpha + 1.618\beta$$

and the bond orders (strengths) determined from the eigenvectors. The complete meaning of these values has been discussed in a number of publications (4,5) and need not be further elaborated on here.

PROBLEMS

1 Modify the main program of SIMULSOLVE to check for zero diagonal elements, and interchange rows until none occur.

2 The *transpose* of a matrix is defined as that matrix obtained by interchanging the rows and columns of a given matrix. If A is a matrix, then A^T is its transpose. If the diagonal matrix D can be obtained by diagonalizing A and the matrix of eigenvectors is called S, then

$A^* = SDS^T$

where A^* should be the same as the original input matrix A. Write a program to calculate A^* and compare it with the original A.

REFERENCES

1 K. Jeffrey Johnson, "Numerical Methods in Chemistry," Marcel Dekker, New York, 1980.

2 T. R. Dickson, "The Computer and Chemistry," Freeman, San Francisco, 1968.

3 A. Ralston and H. S. Wilf, "Mathematical Methods for Digital Computers," vol. 1, Wiley, New York, 1960.

4 J. D. Roberts, "Notes on Molecular Orbital Calculations," Benjamin, New York, 1961.

5 J. W. Cooper, "Spectroscopic Techniques for Organic Chemists," Wiley-Interscience, New York, 1980.

CHAPTER 18

Simplex Optimization of Variables

In many phases of scientific work it is necessary to adjust a series of variables to achieve the best possible response. This can occur, for example, in adjusting various interacting knobs on some instrument or in fitting experimental data to a theoretical function. In such optimization procedures it is necessary to vary the various parameters in some uniform manner to achieve the best possible response.

The simplest optimization procedure starts with some guessed set of parameters and then slowly varies each of them until some best response is achieved. Then it varies the next in a similar fashion. This procedure has the disadvantage of being very slow and inefficient and somewhat likely to find local maxima that do not represent the overall best response. Programs for such optimization procedures have been given in FORTRAN by Bevington (1).

A more efficient method of optimizing these variables is given by the *simplex* approach, originally described by Spendley, Hext, and Himsworth (2) and later applied to chemical parameters by Long (3). Reviews of the simplex method have been given by Deming and Morgan (4) and by Shavers, Parsons, and Deming (5).

In order to discuss this method without confusion, it is necessary to decide what a "best" response means. In optimization of an instrument, it may be a maximum response, while in fitting a function to an equation, it usually means a minimum of the sums of the squares of the differences between the experimental data and the calculated data. In this treatment we will assume that a "better" response has a *smaller* value, as in this least-squares minimization process.

THE SIMPLEX METHOD

A simplex is merely a figure having $N + 1$ vertices or corners in N dimensions, where N is the number of variables to be optimized. If two variables are to be optimized, the simplex is a three-cornered or triangular figure in two-dimensional space. The response at each of these vertices is measured or calculated and a new simplex generated by eliminating the worst point of the original simplex and calculating the position of a new point by reflecting around the center defined by the remaining points. Thus the new point is always as far as possible from the worst response observed. Then the points of this new simplex are evaluated, the new worst one is eliminated, and further simplexes are obtained until the response cannot be improved further.

While simplexes can have any number of vertices in any number of dimensions, it is convenient to illustrate the process graphically by using a simple two-dimensional three-vertex simplex. In such a case we will have only two variables to optimize, and we plot one variable versus the other in two dimensions to represent the three points at which we will measure the response. We then rank these three points as *B, N,* and *W* for best, next, and worst. The first simplex we obtain looks like this:

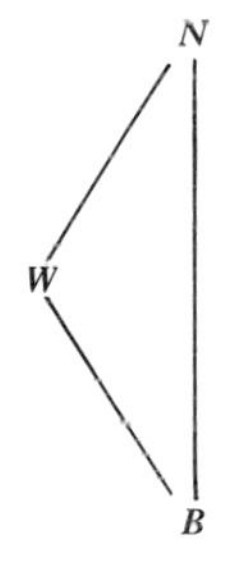

The simplex procedure then calculates the new point to measure, *P,* by reflecting away from the worst response *W* by simple coordinate geometry. This calculation is obvious for this simple two-dimensional case, but it can be expanded to any number of variables in any number of dimensions.

First, the average coordinates (or coordinates of the *centroid*) are calculated for all points except the worst one P_i:

$$C = \frac{1}{k}\ (P_1 + P_2 + \ldots + \mathrm{P}_{i-1} + \mathrm{P}_{i+1} + \ldots + P_n + P_{n+1}) \quad (18.1)$$

where C is a point having coordinates in all N dimensions which are the average coordinates of all but the worst point. Then the coordinates of the new point are calculated by reflecting away from the worst point around the centroid point:

$$P_{\text{new}} = C + (C - W) \quad (18.2)$$

In the diagram below, the centroid of all points but the worst, *W,* is represented by *C,* and the distance *WC* is reflected to form the new point *P* at a distance *WC* on the other side of the centroid. This also forms a new simplex *BNP.*

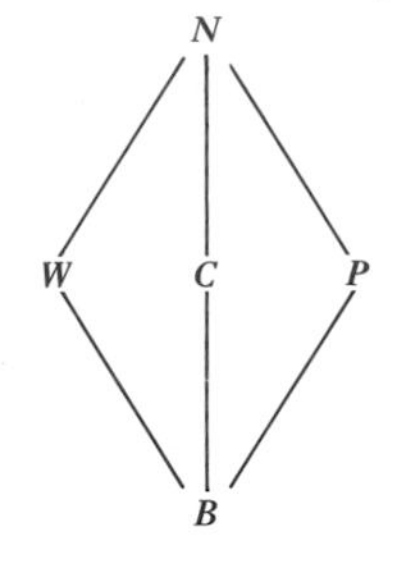

We then move on to the next simplex by evaluating the order of *B*, *N*, and *W* in this second simplex, eliminating the worst point, and reflecting about a new centroid, resulting in:

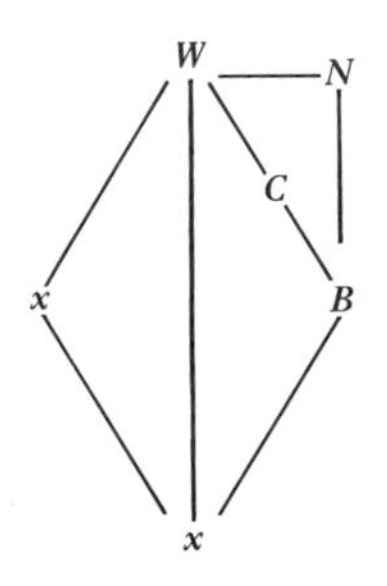

We need only one additional rule for the simplex procedure, to account for the case where the response at the new point is the worst point of the new simplex, as is the case for the simplex BNW':

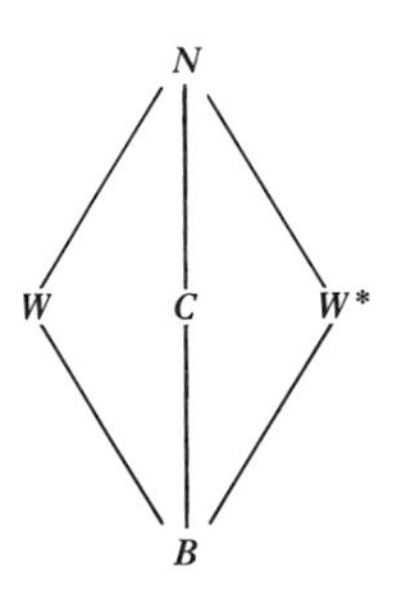

In this case the rules would reflect back around C to give the old simplex BNW back again. This is a case of a simplex being "stranded." Instead, we eliminate not W' of the new simplex, but the next worst point N instead, and reflect across BW' to give the newest simplex.

In summary, we give the following rules for the original simplex procedure (4):

1 A move is made after the observation of each response.

2 The above is made into that adjacent simplex which is obtained by discarding the point of the current simplex consisting of the least desirable response and replacing it with its mirror image across the centroid of the remaining points.

3 If the reflected point has the least desirable response of the new simplex, then avoid reflecting back to the previous worst point by discarding the next worst point instead.

4 If the new vertex of the simplex lines outside the bounds of reasonable or legal values for one or more variables, return a very undesirable response for that measurement.

DISADVANTAGES OF THE ORIGINAL SIMPLEX PROCEDURE

This original simplex procedure has the main disadvantage that it can never achieve a finer measurement distance between points than was originally specified as the distance between the first $N + 1$ points. This was solved by simply allowing the simplex to cycle until the best response possible was achieved, and then running the simplex program again with a smaller spacing between the points of the last simplex found. Further, if the original choices of points are too close together, there is no provision for acceleration to arrive at the optimum faster.

In addition, in cases of more than two dimensions, the simplex will not necessarily home in on the top of the desired response, but will circle it rather inefficiently. This makes it difficult to know when to discontinue the operation and reduce the distance between points.

THE MODIFIED SIMPLEX

The modified simplex procedure grew out of these disadvantages and has been discussed by Nelder and Mead (6) and reviewed by Deming and co-workers (4, 5). In this procedure the distance between points can be expanded or contracted, depending on the relationship between the points of each simplex:

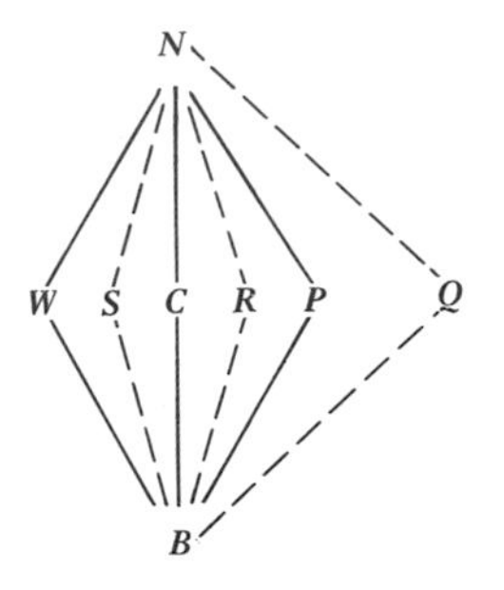

After reflecting to find the new point P in the modified simplex procedure, we examine additional points near P to accelerate or decelerate the simplex:

1. If P is the best response of the new simplex, expand the simplex in that direction by doubling the reflection distance CP to $2CP$. This will produce the new point Q. If Q is better than P, replace it as the new vertex, and if it is worse, leave P as the new simplex point.
2. If $W < P < N$, shorten the reflection axis CP to half as long, CR. Replace P with R if R is the better response. (Here we use $<$ to mean W "is worse than" since we may be maximizing or minimizing.)
3. If $P < W$, measure the response S halfway between W and C and replace P with that point if it is better.
4. If none of the above are true, leave P as the new point of the simplex.

In summary, for the modified simplex procedure we modify Eq. 18.2 to give:

$$P_{\text{new}} = C + F(C - W) \tag{18.3}$$

where the factor F is determined by the evaluation of P with respect to B, N, and W:

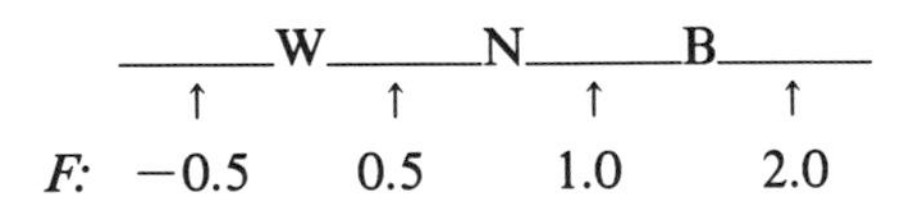

THE PROGRAM FOR MODIFIED SIMPLEX OPTIMIZATION

The following program uses the modified simplex procedure to fit a set of data points to a function. The function is shown in FUNC and can be changed at will. The program starts by reading in the number of point pairs and then reads in the arrays of X's and Y's which are the experimental data. Then it reads in the number of variables in the equation, the original guessed values of each of the variables, and the amount to vary them to generate the first simplex.

It uses the least-squares criterion to minimize the sum of the squares of the differences between the experimental points and the calculated ones. Lower case comments are shown to illustrate their readability.

```
PROGRAM SIMPLEXDEMO;

CONST VARIABLES=10; (*must be 2 greater than max number of variables*)

TYPE

     VINDEX=1..VARIABLES;
     VARAY=ARRAY[VINDEX] OF REAL;
     VMAT=ARRAY[VINDEX,VINDEX] OF REAL;
VAR
A:VMAT; (*Array of variables vs measurements*)
X,Y:ARRAY[1..100] OF REAL;  (*Array of points on curve*)
CENTROID,DELTA,RESPONSE:VARAY;
NPTS,VMAX,VMAXP1,MEAS,V,I,BEST,WORST,
NEXTBEST,WORST2:VINDEX;
ITER,NEW2,NEWPNT:INTEGER;
TOL:REAL;    (*tolerance below which chisq must fall*)
F1: TEXT;    (*input data file*)

(*********************************************************)
FUNCTION FUNC(K,L:INTEGER):REAL;
(*This function evaluates the calculated value of Y
for a given X[L] and uses the values of the
coefficients in the Kth column of A*)
BEGIN
     FUNC:=A[1,K] + A[2,K]*EXP(X[L]*A[3,K]);
END;(*FUNC*)
(*********************************************************)
PROCEDURE MEASURE(M:INTEGER);

VAR
CALCY,CHISQR:REAL;
I:INTEGER;

(*This procedure uses the mth values of the variables in A
to calculate a new response(m). The response is simply the
chi square value obtained by plugging the values of the
a's into the equation for each I and calculating
a CALCY which is compared with the observed Y[I].
The square of the resulting difference is added to CHISQR, which
becomes the response when done*)

BEGIN
ITER:=ITER+1;     (*COUNT MEASUREMENTS*)
CHISQR:=0.0;      (*Chi square accumulates here *)
FOR I:=1 TO NPTS DO
     BEGIN
     (*Here the function is evaluated for the mth column
     of A and the ith value of X*)
     CALCY:=FUNC(M,I);
     CHISQR:=CHISQR + SQR(Y[I]-CALCY);
     END;
 RESPONSE[M]:=CHISQR;     (*Return chi square value as response*)
 END; (*MEASURE*)
```

```
(****************************************************************)
PROCEDURE SETUP;
(*Make the first VMAX+1 measurements on all VMAX variables*)

VAR V,MEAS:VINDEX;

BEGIN
VMAXP1:=VMAX+1;
FOR MEAS :=1 TO VMAXP1 DO
     BEGIN
     MEASURE(MEAS);
     (*Calculate the next parameters needed
     from the values of DELTA. Switch their signs so
     that they are not all in a straight line *)

     FOR V :=1 TO VMAX DO
     BEGIN
     IF V=MEAS THEN DELTA[V]:=-DELTA[V];
     A[V,MEAS+1]:=A[V,MEAS] + DELTA[V];
     END;(*FOR V*)
     END;(*FOR MEAS*)
END;(*SETUP*)
(****************************************************************)
PROCEDURE MOVE(L,M:VINDEX);
(*Puts the variables in column M in column L of A*)

VAR V:VINDEX;

BEGIN
     FOR V:=1 TO VMAX DO A[V,L]:=A[V,M];
     RESPONSE[L]:=RESPONSE[M];
END;     (*MOVE*)

(****************************************************************)
PROCEDURE GENNEW(M:VINDEX;F:REAL);
(*Generates new vertices for the simplex by
multiplying the difference between the worst and the
centroid by the factor F, and then measuring the new
value. Keeps the new value only if better than the
original one*)

VAR V:VINDEX;

BEGIN
FOR V:=1 TO VMAX DO
    A[V,M]:=CENTROID[V]+F*(CENTROID[V]-A[V,WORST]);
MEASURE(M);

(*Replace worst value with new one if this response
is better and if this is not the initial measurement
with the factor of 1.00*)
IF F<>1.0 THEN
    IF RESPONSE[M]<RESPONSE[NEWPNT] THEN MOVE(WORST,M)
    ELSE MOVE(WORST,NEWPNT);
END;     (*GENNEW*)
(****************************************************************)

PROCEDURE MODSIMPLEX;
(*Simplex optimization procedure, operates on VMAX different
variables in a matrix A(variables,measurements)
where there must always be one more measurement than the total number
of parameters to be varied. Since we are minimizing
the least squares response, BEST is the smallest response
and WORST the largest*)
VAR
MAX,MIN,MAX2,MIN2,ROLD:REAL;
I,V:VINDEX;
```

```
BEGIN
NEWPNT:=0;  (*This is the index of the newest measured point*)
VMAXP1:=VMAX+1; (*number of measurements needed altogether*)

NEWPNT:=VMAX+2; (*Extra column for new measurement*)
NEW2:=VMAX+3;   (*Extra column for 2nd new measurement*)
REPEAT
(*Find the maximum and minimum response measured *)
MAX:=0.0;    MIN:=1.0E36;
FOR I:=1 TO VMAXP1 DO
     BEGIN
     IF RESPONSE[I]>MAX THEN
          BEGIN
          MAX:=RESPONSE[I];    (*Make new MAX*)
          WORST:=I;
          END;
     IF RESPONSE[I]<MIN THEN
          BEGIN
          MIN:=RESPONSE[I];    (*Save new MIN *)
          BEST:=I;
          END;
     END;     (*FOR I*)

(*Find 2nd best and 2nd worst, too*)
MAX2:=0;     MIN2:=1.0E36;
FOR I := 1 TO VMAXP1 DO
     BEGIN
     IF (RESPONSE[I]>MAX2) AND (RESPONSE[I]<MAX) THEN
          BEGIN
          MAX2:=RESPONSE[I];
          WORST2:=I;
          END;(*IF*)
     IF (RESPONSE[I]<MIN2) AND (RESPONSE[I]>MIN) THEN
          BEGIN
          MIN2:=RESPONSE[I];
          NEXTBEST:=I;
          END;(*IF*)
     END;(*FOR I*)
(*Calculate centroid of all measurements except worst.*)
FOR I:=1 TO VMAX DO
     CENTROID[I]:=0; (*Zero out centroids first*)
FOR V:=1 TO VMAX DO
     BEGIN
     FOR MEAS:=1 TO VMAXP1 DO
     IF MEAS<>WORST THEN
     CENTROID[V]:=CENTROID[V] + A[V,MEAS];
     END;(*FOR*)
(*Divide centroid down by VMAX *)
FOR I:= 1 TO VMAX DO
     CENTROID[I]:=CENTROID[I]/VMAX;

(*Measure the response at the point reflected away from worst*)
GENNEW(NEWPNT,1.0); (*Measure new point normally*)

(*If this one is better than previous best, then expand in this direction*)
IF RESPONSE[NEWPNT]<RESPONSE[BEST] THEN
     GENNEW(NEW2,2.0)

(*If this one is worse than previous worst, measure
point halfway between worst and centroid*)
ELSE IF RESPONSE[NEWPNT]>RESPONSE[WORST] THEN
     GENNEW(NEW2,-0.5)
```

```
(*If newest response is worse than next best point
but better than worst, measure response halfway
between centroid and newest point*)
ELSE IF (RESPONSE[NEXTBEST]<RESPONSE[NEWPNT]) AND
    (RESPONSE[NEWPNT]<RESPONSE[WORST]) THEN
    GENNEW(NEW2,0.5)

(*If none of the above, keep the new point as best*)
ELSE MOVE(WORST,NEWPNT);
UNTIL RESPONSE[NEWPNT]<TOL;
END;      (*MODSIMPLEX*)

(*************************************************************)
BEGIN    (*MAIN*)

RESET(F1,'SIMINP    ');   (*Open file SIMINP for input data*)
WRITE('TOLERANCE=');      (*set allowed tolerance from terminal*)
READLN(TOL);
ITER:=0;
(*Read in number of points to fit*)
READLN(F1,NPTS);

(*Then read in x,y pairs*)
FOR I:=1 TO NPTS DO READLN(F1,X[I],Y[I]);

(*Read in # of variables, initial guess for parameters and deltas*)
READLN(F1,VMAX);      (*Number of variables*)
FOR I:=1 TO VMAX DO READLN(F1,A[I,1],DELTA[I]);

(*Make N+1 measurements to set up the A[variable,measurement] matrix*)
SETUP;

(*Now perform simplex optimization*)
    MODSIMPLEX; (*Call the simplex procedure*)

FOR I:=1 TO VMAX DO WRITE(A[I,NEWPNT]);
WRITELN(RESPONSE[NEWPNT]);
WRITELN(ITER,' ITERATIONS');
END.
```

DESCRIPTION OF THE PROCEDURES IN SIMPLEXDEMO

The overall strategy of this simplex procedure is to collect VMAX+1 measurements of the VMAX variables to be optimized in a matrix A whose columns are the variables and whose rows are the measurements. Two additional columns are also provided for additional measurements to be made and compared with the last best measurement. Thus the maximum dimensions of the array are actually VMAX by VMAX+3, where there are VMAX+1 columns of VMAX variables plus two extra columns for alternative measurements NEWPNT and NEW2.

MEASURE

The measure routine uses the values of the variables in column m of the matrix A to evaluate the response of the function. The response is the sum of the squares of the differences between the measured Y points at given X values and the calculated Y points. This is accumulated in CHISQ and returned in the mth element of the array RESPONSE.

FUNC

The FUNC routine defines the actual function to be fitted to the curve. There are three variables to be optimized in this case to the function:

$$y_i = a_1 + a_2 \exp(a_3 x_i)$$

SETUP

The SETUP routine uses the initial guessed values for the parameters to be varied and the values of the steps DELTA to calculate the vertices of the first simplex and measure the response at these vertices. So that the points will not lie in a straight line, the procedure negates the Vth delta when the Vth measurement is taken.

MOVE

Since a decision is made about whether to replace a given simplex vertex with a new measurement based on the value of the response at that vertex compared to existing vertices, this routine moves the new values of the variables and response into the actual array from one of the two extra columns NEWPNT and NEW2.

GENNEW

The GENNEW procedure calculates the new values of the variables from Eq. 18.3 where the value of F is passed to the procedure from the MODSIMPLEX program. The new measurement replaces the one in the WORST vertex of the current simplex, unless this is the first measurement before decisions of replacement are made, when F = 1.0.

MODSIMPLEX

This is the modified simplex routine. It finds the best and worst measurements and the next best, calculates the coordiates of the centroid of all vertices except that having the worst response, and makes the measurement of the predicted new vertex. Then it measures another point, depending on whether the value of the new point's response is better than the previous best, worse than the previous worst, worse than the next best point, or between the nextbest and the best. The procedure continues until a response $<$ TOL is found.

MAIN

The main driver program reads in the number of experimental point pairs, the point pairs, the number of variables, and the original guessed values for the variables and for the steps to take to the other measurements making up the original simplex. It also reads in a tolerance value TOL which is the value below which chi square must fall before the program will exit.

RESULTS OF THE SIMPLEX PROGRAM

For the input file having eight measured points and the original values of $a_1, \ldots, a_3$ of 1, 1, 1 and deltas of 0.1, 0.1, and 0.1:

```
8
1.0  2.80
2.0  3.74
3.0  5.01
4.0  6.74
5.0  9.06
6.0  12.20
7.0  16.43
8.0  22.14

3
1.0  −0.1
1.0  −0.1
1.0  0.1
```

and for a value of TOL of 0.0001, the program produces values of

$$a_1 = 0.1000888$$
$$a_2 = 1.999029$$
$$a_3 = 0.300044$$

and a chi square of

$$8.76718E\text{-}05$$

in 187 iterations.

REFERENCES

1 P. R. Bevington, "Data Reduction and Error Analysis for the Physical Sciences," McGraw-Hill, New York, 1969.

2 W. Spendley, G. R. Hext, and F. R. Himsworth, *Technometrics,* 4, 441 (1962).

3 D. E. Long, *Anal. Chim. Acta,* 46, 193 (1969).

4 S. N. Deming and S. L. Morgan, *Anal. Chem.,* 45, 278A (1973).

5 C. L. Shavers, M. L. Parsons, and S. N. Deming, *J. Chem. Educ.,* 56, 307 (1979).

6 J. A. Nelder and R. Mead, *Comput. J.,* 7, 308 (1965).

CHAPTER 19

The Fourier Transform

Uses of the Fourier Transform
Development of the Transform
The Cooley-Tukey Algorithm
The Signal Flow Graph
Bit-Inverted Order
The Form of *W*
Fourier Transforms of Real Data
Saving Calculation Time in Fourier Transforms
A Fourier Transform Routine
Example of Fourier Transform Data
Problems
References

USES OF THE FOURIER TRANSFORM

In many applications in science, data is acquired from some sort of instrument as a series of sine waves. Such waves may be the harmonics of a voice, used in speech analysis, or data from infrared interferograms (1), nuclear magnetic resonance data (2, 3), or results of vibrational studies. In every case the combinations of sine waves can be difficult to interpret visually, and the Fourier transform is used to convert them to peaks. For example, consider the single decaying sine wave and peak below:

It is easy to see from either plot that there is only a single frequency here and conversion of the data could be made by inspection. We define the plot of the sine wave as a *time domain* plot since it usually represents a plot of absorption versus time. The peak representation is a *frequency domain* plot since it represents a plot of absorption versus frequency.

In the second figure below

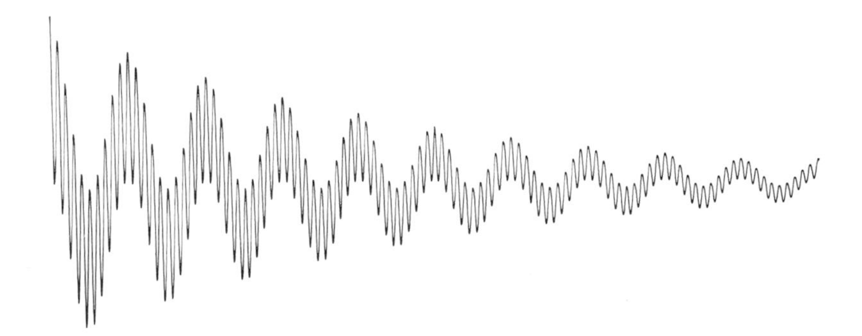

we see that there are two co-added sine waves, and again we can easily distinguish them in either the time domain or the frequency domain plot.

Finally, in the third figure below

we see that there are so many sine waves co-added that we cannot count them all, and thus could not make the frequency domain plot by inspection. Instead, this conversion to the frequency domain is made by the Fourier transform process.

The Fourier transform, then, is a mathematical routine for the conversion of time domain to frequency domain data *and vice versa.* The conversion from time to frequency is usually called a *forward transform,* and the conversion from frequency back to time is usually called an *inverse transform.*

DEVELOPMENT OF THE TRANSFORM

The actual discrete digital expression for the Fourier transform is given by the Eq. 19.1:

$$A_r = \sum_{k=0}^{N-1} (X_k) W^{rk}, \qquad r = 0, \ldots, N-1 \tag{19.1}$$

where

$$W = \exp\left(\frac{-2\pi i}{N}\right) \tag{19.2}$$

In this equation X_k is the kth time domain point and A_r is the rth frequency domain point. The X's may be complex numbers and the A's always are complex numbers. Since it clearly takes N multiplications to calculate one A_r, it must take N^2 multiplications to calculate all N A_r's. Since multiplication is one of the slowest operations a computer performs, this calculation can be very costly in computer time, and was the bottleneck in the data reduction process until the Cooley-Tukey algorithm was published in 1965.

THE COOLEY-TUKEY ALGORITHM

The Cooley-Tukey algorithm (4), or the *method* for calculating Fourier transforms, relies on the ability to factor the data into sparse matrices containing mostly zeros. A good matrix algebra proof of the transform has been given by Brigham (5), and thorough proofs of the ability to factor the data into smaller transforms have been given in several places (6–8).

To understand why the Cooley-Tukey algorithm works, let us suppose that we divide the time domain array X_k into two smaller arrays Y_k and Z_k consisting of the even and odd points of the original array, and that these have the transforms B_r and C_r, respectively. Then it can be shown that we can obtain A_r, the transform of X_k, by combining the transforms B_r and C_r:

$$\begin{aligned} A_r &= B_r + W^r C_r \\ & \qquad\qquad\qquad 0 \leq r < \frac{N}{2} \\ A_{r+n/2} &= B_r - W^r C_r \end{aligned}$$

This is important because while the transformation of N points requires N^2 multiplications, it only requires

$$2\left(\frac{N}{2}\right)^2 = \frac{N^2}{2}$$

multiplications to transform the two smaller arrays.

In a similar fashion, we could further subdivide these Y_k and Z_k arrays and obtain their transforms, down to the case where each sub-sub- . . . -sub array consists of only one point. The transform of a single point is the point itself, and we have thus simplified the transform to the case where we do not actually do a classical transform at all. Instead, we only need to perform the recombination arithmetic to arrive at the actual array in many less multiplications. In fact, it turns out that while a classical transform requires N^2 multiplications, the Cooley-Tukey transform procedure requires only $N \log_2 N$ multiplications. For a typical small array of, say, 4096 points, the classical transform would require $(4096)^2$ or 16.7 million multiplications, while the Cooley-Tukey algorithm requires only

$$4096 \log_2(4096) = 4096(12) = 49{,}152$$

multiplications, saving a factor of 341 in time. It is a requirement, however, that the array to be transformed contain a power of 2 number of points.

THE SIGNAL FLOW GRAPH

An easy way to understand the Fourier transform calculation process is through the use of a signal flow graph, such as that given on p. 204.

This graph shows the transform of eight *complex* points from the time domain (X's) to the frequency domain (A's). Each intersection at a dot represents the complex addition of two points to form a new point, and each W term alongside a line represents complex multiplication of that point by that power of W before the addition takes place. Thus each new point at a dot or *node* is calculated by:

$$X_i' = X_i + W^y X_k$$

where X_i' is the new point, y the power of W at that point, and X_k another X point with which it is combined.

Looking further into the graph, we see that there are several intermediate columns of points calculated between the original X's and the final A's. Each of these columns is called a *pass* and represents an intermediate result in the transform process. There will always be $\log_2 N$ passes through the transform for N complex points.

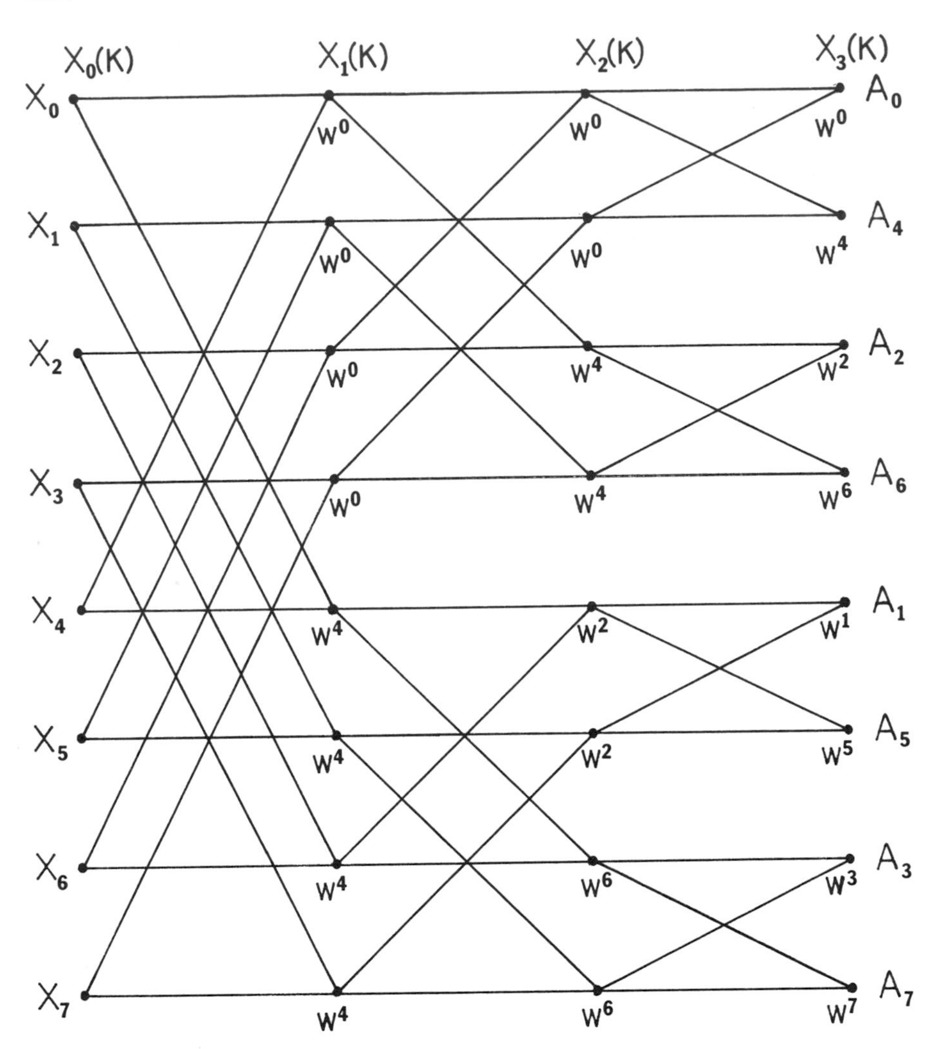

Within each pass we see that only two points go to make up each new point and that these two points influence only two points in the new pass. The two equations describing these two points have the form:

$$X_1' = X_1 + X_2W^y, \qquad X_2' = X_1 + X_2W^z \tag{19.3}$$

and thus we can carry out this transform *in place* since we operate on points a pair at a time.

BIT-INVERTED ORDER

If we examine the order of the A's and X's in the graph, we see that while the X's are presented in order, the A's are in an unusual order: 0, 4, 2, 6, 1, 5, 3, 7. This strange order can be decoded by writing the numbers down in binary:

```
0  000
4  100
2  010
6  110
1  001
5  101
3  011
7  111
```

and comparing them with the numbers in binary in their natural order:

```
0  000
1  001
2  010
3  011
4  100
5  101
6  110
7  111
```

We see that the order in which the A's finish up in is predicted by simply reading their binary indices from right to left rather than from left to right. This order is called *bit-reversed* or *bit-inverted* order, and the points are easily shuffled in and out of this order.

As we see in the figure below, the Fourier transform can also be performed by starting with the X indices in bit-inverted order and ending up with the A indices in natural order. This second flow chart, with bit inversion first, has the additional advantage that the orders of the W exponents are easily discernable and thus easier to program. We will be using this approach in the transform in this chapter.

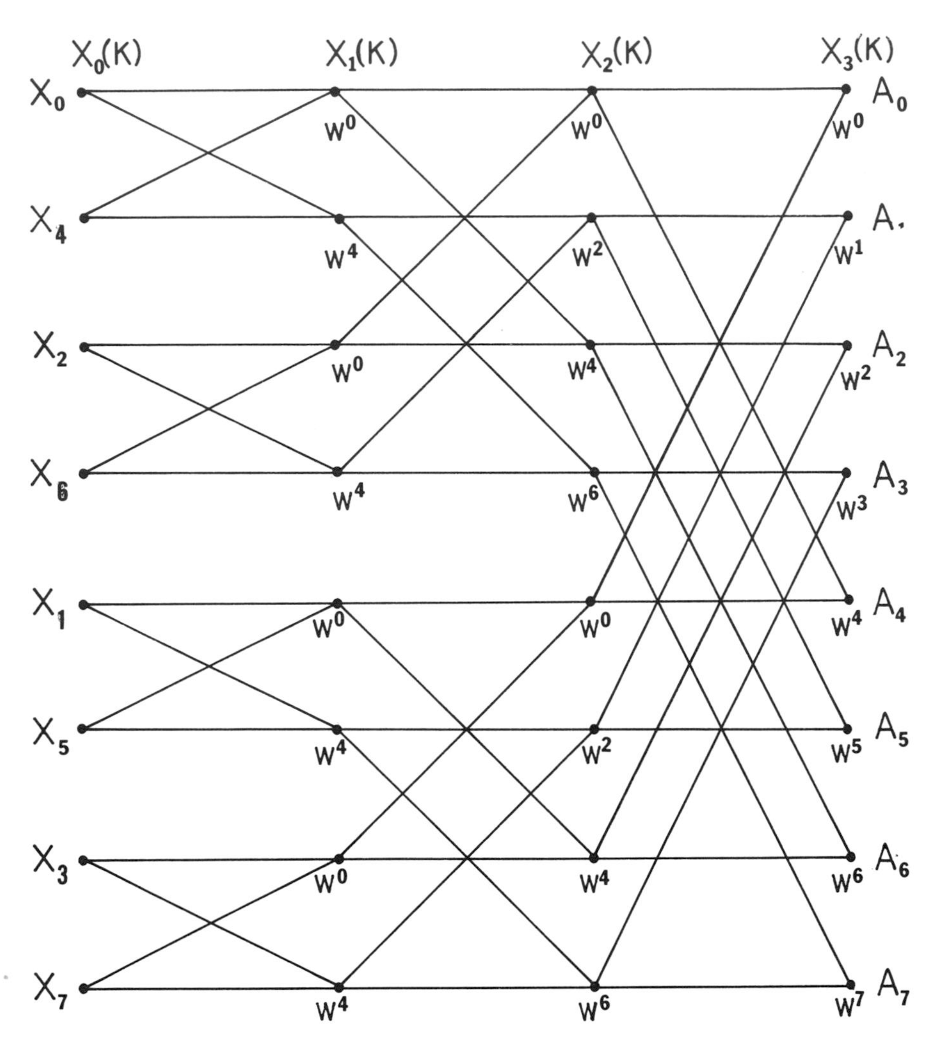

THE FORM OF W

We have defined the term W as

$$W = \exp\left(\frac{-2\pi i}{N}\right) \tag{19.2}$$

However, this particular form is quite time consuming to use, since the calculation of a number of exponentials can be very slow and awkward in the computer. Instead, we invoke Euler's formula:

$$e^{iy} = \cos(y) + i\sin(y) \tag{19.4}$$

and then represent W as

$$W^y = \exp\left(\frac{-2\pi iy}{N}\right) - \cos\left(\frac{-2\pi y}{N}\right) + i\sin\left(\frac{-2\pi y}{N}\right)$$

For $y = 3$, then, we have

$$W^3 = \exp\left(\frac{6\pi i}{8}\right) = \cos(0.75\pi) + i\sin(0.75\pi) = -0.707 + 0.707i$$

Since all of the X's are complex, we have multiplication of the complex number X by the complex number derived from W. We further note that the exponents of W for $X1'$ and $X2'$, which we term y and z above, are related in that they differ by $N/2$:

$$z = y + \frac{N}{2} \tag{19.5}$$

and thus that:

$$W^z = W^{y+N/2} = W^y \exp\left(\frac{-2\pi iN}{2N}\right) = W^y(-1)$$

Thus we have:

$$W^z = -W^y$$

The two Eqs. 19.3 that we evaluate for any pair of complex points can then be written as:

$$\begin{aligned} X_1' &= X_1 + X_2W^y \\ X_2' &= X_1 - X_2W^y \end{aligned} \tag{19.6}$$

Now if we expand the representations of X to represent the real and imaginary coefficients,

$$X_1 = R_1 + iI_1, \qquad X_2 = R_2 + iI_2$$

then we can expand Eq. 19.6 to:

$$\begin{aligned} X_1' &= R_1' + iI_1' = (R_1 + iI_1) + (R_2 + iI_2)(\cos y + i\sin y) \\ X_2' &- R_2' \mid iI_2' = (R_1 + iI_1) - (R_2 + iI_2)(\cos y + i\sin y) \end{aligned} \tag{19.7}$$

Collecting terms, we have:

$$
\begin{aligned}
R_1' &= R_1 + R_2 \cos y - I_2 \sin y \\
R_2' &= R_1 - R_2 \cos y + I_2 \sin y \\
I_1' &= I_1 + R_2 \sin y + I_2 \cos y \\
I_2' &= I_1 - R_2 \sin y - I_2 \cos y
\end{aligned}
\tag{19.8}
$$

These are the fundamental equations of the Fourier transform, and they are evaluated for all complex points for a given set of y's to form one *pass* through the data. Then the size of the increment to y is decreased and the process repeated again, until the size of the "cell" in which y goes through one cycle is the same as the size of the array. Then the process is complete.

FOURIER TRANSFORMS OF REAL DATA

Thus far we have assumed that the transform will be of complex data. Much scientific data is, in fact, all real, and the process for transforming it is of some concern. The following method is that given by Brigham(5).

1 The data is divided into odd and even points which are arbitrarily called the real and imaginary points for the purposes of the transform.

2 The data is shuffled so that the real part is in the first half of the array and the imaginary in the second half of the array.

3 The complex transform of these data is computed.

4 The transformed function is B_r, consisting of a real and an imaginary part. We convert this to the desired array A_r by applying the simple one-pass post-processing algorithm:

$$
\left.
\begin{aligned}
A_r(n) &= R_p + \cos \frac{\pi n}{N} I_p - \sin \frac{\pi n}{N} R_m \\
A_i(n) &= I_m - \sin \frac{\pi n}{N} I_p - \cos \frac{\pi n}{N} R_m
\end{aligned}
\right\} \quad n = 0, 1, \ldots, N-1
$$

where

$$R_p = R_n + R_{M-n}$$
$$I_p = I_n + I_{M-n}$$
$$R_m = R_n - R_{M-n}$$
$$I_m = I_n - I_{M-n}$$

and

$$M = \frac{N}{2}$$

While the evaluation of the equation generates $N/2$ complex points, it requires $N/2 + 1$ input points. The symmetry of the transform allows us to calculate the first and last points separately:

$$R_p = R_0$$
$$I_p = 0$$
$$R_m = 0$$
$$I_m = -R_0$$

SAVING CALCULATION TIME IN FOURIER TRANSFORMS

There are several ways to further speed up the Fourier transform, notably the elimination of multiplications whenever the sine and cosine functions have predictable values, such as the sine and cosine of 0.

In the first pass through the transform, the powers of W are only 0 and $N/2$, and thus the fundamental equations reduce to additions and subtractions only:

$$X1' = X1 + X2$$
$$X2' = X1 - X2$$

Since the powers of W always start at 0 in each cell, it makes sense in large transforms to trap for the 0 sine case and treat it separately. This leads to the same equations as above for the first point pair in each cell.

In minicomputer implementations it is also useful to provide a sine *look-up* table to avoid the necessity of calculating each sine and cosine through a Taylor series (7, 8).

The usual implementation of the Fourier transform has the form:

```
set DELTAY to π/2
set CELLCOUNT to N/2
set POINTCOUNT to 2
REPEAT (*UNTIL DELTAY = 0*)
   REPEAT (*CELL COUNT TIMES*)
   set Y=0
      REPEAT
      X_i = X_i + W^y X_j
      X_j = X_i - W^y X_j
      POINTCOUNT times
   CELLCOUNT times
CELLCOUNT = CELLCOUNT/2
POINTCOUNT=POINTCOUNT * 2
DELTAY=DELTAY/2
UNTIL DELTAY= 0
```

We have noted (9) that an additional 20 percent in speed can be obtained by doing all the first points in each of the cells of the pass, then all of the second points, and so forth, since this obviates the necessity of recalculating the sines and cosines for each cell of the pass. This is outlined below:

```
set DELTAY to π/2
set CELLCOUNT to N/2
set POINTCOUNT to 2
set CELLDISTANCE to 2
REPEAT
set Y=0
   REPEAT
      REPEAT
      X_i = X_i + W^y X_j
      X_j = X_i - W^y X_j
      I= I+CELLDISTANCE
      J= J+CELLDISTANCE
      CELLCOUNT times
   POINTCOUNT times
   CELLCOUNT=CELLCOUNT/2
   POINTCOUNT= POINTCOUNT*2
   DELTAY=DELTAY/2
UNTIL DELTAY=0
```

where we simply interchange the middle and outer loops. The program that follows uses all of these features to produce the most time-efficient transform possible.

A FOURIER TRANSFORM ROUTINE

The program that follows contains a complete set of procedures for the calculation of the Fourier transform. As it is written, the calls produce a forward transform of all real data, but the other possibilities are also available by different calls to the routines:

```
    (*COMPLEX TRANSFORM-1ST HALF REAL, 2ND HALF IMAG*)
    FFT(FWD)

    (*COMPLEX INVERSE*)
    FFT(INVERSE);

    (*REAL FORWARD TRANSFORM*)
    SHUFFL(FWD);
    FFT(FWD);
    POST(FWD);

    (*REAL INVERSE TRANSFORM*)
    POST(INVERSE);
    FFT(INVERSE);
    SHUFFL(INVERSE);     (*PUT BACK INTO ALTERNATE ORDER*)

    (*COMPLEX INVERSE WHERE FREQ PTS ARE REAL 1ST, IMAG 2ND HALF*)
    FFT(INVERSE);
    SHUFFL(INVERSE);

PROGRAM FOURIER;
(*REAL FFT WITH SINGLE SINE LOOK-UP PER PASS*)
CONST
    ASIZE=4096;          (*SIZE OF ARRAY GOES HERE*)
    PI2=1.570796327;     (*PI OVER 2*)
    F=16;       D=2;     (*FORMAT CONSTANTS*)
TYPE
    XFORM=(FWD,INVERSE);     (*TRANSFORM TYPES*)
    XARY=ARRAY[1..ASIZE] OF REAL;

VAR
    I,N:INTEGER;
    F1:FILE OF REAL;     (*DATA FILE OF REAL NUMBERS*)
    X:XARY; (*ARRAY TO TRANSFORM*)
    INV:XFORM;  (*TRANSFORM TYPE-- FORWARD OR INVERSE*)

(*************************************************************************)
PROCEDURE DEBUG;     (*USED TO PRINT OUT INTERMEDIATE RESULTS*)

VAR I3:INTEGER;

BEGIN
```

```
     FOR I3:=1 TO N DO
     WRITELN(X[I3]:F:D);
END;     (*DEBUG*)
(**************************************************************************)
FUNCTION IBITR(J,NU:INTEGER):INTEGER;
(*FUNCTION TO BIT INVERT THE NUMBER J BY NU BITS*)

VAR
     I,J2,IB:INTEGER;

BEGIN
     IB:=0;  (*DEFAULT RETURN VALUE*)
     FOR I:=1 TO NU DO
          BEGIN
          J2:=J DIV 2;     (*DIVIDE BY 2 AND COMPARE LOWEST BITS*)
          (*IB IS DOUBLED AND BIT 0 SET TO 1 IF J IS ODD*)
          IB:=IB*2 + (J- 2*J2);
          J:=J2;  (*FOR NEXT PASS*)
          END;     (*FOR*)
     IBITR:=IB;  (*RETURN BIT INVERTED VALUE*)
END;     (*IBITR*)
(**************************************************************************)
PROCEDURE POST(INV:XFORM);
(*POST PROCESSING FOR FORWARD REAL TRANSFORMS AND
PRE-PROCESSING FOR INVERSE REAL TRANSFORMS, DEPENDING
ON STATE OF THE VARIABLE INV*)

VAR
     NN2,NN4,L,I,M,IPN2,MPN2:INTEGER;
     ARG,RMSIN,RMCOS,IPCOS,IPSIN,IC,IS1,RP,RM,IP,IM:REAL;

     BEGIN
     NN2:=N DIV 2;    (*N IS GLOBAL!*)
     NN4:=N DIV 4;
(*IMAX REPRESENTS PI/2*)
FOR L:= 1 TO NN4 DO
(*START AT ENDS OF ARRAY AND WORK TOWARDS MIDDLE*)
     BEGIN
     I:=L+1;            (*POINT NEAR BEGINNING*)
     M:=NN2-I+2;        (*POINT NEAR END*)
     IPN2:=I+NN2;           (*AVOIDS RECALCULATION EACH TIME*)
     MPN2:=M+NN2;       (*CALCS PTRS TO IMAGINARY PART*)
     RP:=X[I]+X[M];
     RM:=X[I]-X[M];
     IP:=X[IPN2]+X[MPN2];
     IM:=X[IPN2]-X[MPN2];
(*TAKE COSINE OF PI/2N*)
     ARG:=(PI2/NN4)*(I-1);
     IC:=COS(ARG);
(*COSINE TERM IS MINUS IF INVERSE*)
     IF INV=INVERSE THEN IC:=-IC;
     IS1:=SIN(ARG);
     IPCOS:=IP*IC;    (*AVOID REMULTIPLICATION BELOW*)
     IPSIN:=IP*IS1;
     RMSIN:=RM*IS1;
     RMCOS:=RM*IC;
     X[I]:=RP + IPCOS -RMSIN;     (*COMBINE FOR REAL-1ST PT*)
     X[IPN2]:=IM -IPSIN - RMCOS; (*IMAG-1ST POINT*)
     X[M]:= RP - IPCOS + RMSIN;        (*REAL- LAST PT*)
     X[MPN2]:= -IM -IPSIN -RMCOS;     (*IMAG, LAST PT*)
     END;(*FOR*)
END;(*POST*)
(**************************************************************************)
```

```
PROCEDURE SHUFFL(INV:XFORM);
(*THIS PROCEDURE SHUFFLES POINTS FROM ALTERNATE REAL-IMAGINARY TO
1ST-HALF REAL, 2ND-HALF IMAGNARY IF INV=FWD,
AND REVERSES THE PROCESS IF INV=INVERSE.
ALGORITHM IS MUCH LIKE COOLEY-TUKEY:
    STARTS WITH LARGE CELLS AND WORKS TO SMALLER ONES FOR FWD
    AND STARTS WITH SMALL CELLS AND INCREASE IF INVERSE*)

VAR
    I,J,K,IPCM1,CELDIS,CELNUM,PARNUM:INTEGER;
    XTEMP:REAL;

BEGIN
(*CHOOSE WHETHER TO START WITH LARGE CELLS AND GO DOWN
OR START WITH SMALL CELLS AND INCREASE*)

CASE INV OF

FWD:     BEGIN
    CELDIS:=N DIV 2;     (*DISTANCE BETWEEN CELLS*)
    CELNUM:=1;           (*ONE CELL IN 1ST PASS*)
    PARNUM:=N DIV 4;     (*N/4 PAIRS PER CELL IN 1ST PASS*)
    END;     (*FWD CASE*)
  INVERSE: BEGIN
     CELDIS:=2;            (*CELLS ARE ADJACENT*)
     CELNUM:=N DIV 4;      (*N/4 CELLS IN 1ST PASS*)
     PARNUM:=1;
     END; (*CASE INVERSE*)
  END;     (*CASE*)

  REPEAT        (*UNTIL CELLS LARGE IF FWD OR SMALL IF INVERSE*)
     I:=2;
     FOR J:=1 TO CELNUM DO
     BEGIN
         FOR K:=1 TO PARNUM DO    (*DO ALL PAIRS IN EACH CELL*)
              BEGIN
              XTEMP:=X[I];
              IPCM1:=I+CELDIS-1;  (*INDEX OF OTHER PT*)
              X[I]:=X[IPCM1];
              X[IPCM1]:=XTEMP;
              I:=I+2;
              END;(*FOR K*)
         (*END OF CELL, ADVANCE TO NEXT ONE*)
         I:=I+CELDIS;
     END;     (*FOR J*)

  (*CHANGE VALUES FOR NEXT PASS*)

  CASE INV OF
  FWD:     BEGIN
     CELDIS:=CELDIS DIV 2;    (*DECREASE CELL DISTANCE*)
     CELNUM:=CELNUM * 2;      (*MORE CELLS*)
     PARNUM:=PARNUM DIV 2;    (*LESS PAIRS PER CELL*)
     END;     (* CASE FWD*)

  INVERSE:     BEGIN
     CELDIS:=CELDIS * 2;      (*MORE DISTANCE BETWEEN CELLS*)
     CELNUM:=CELNUM DIV 2;    (*LESS CELLS*)
     PARNUM:=PARNUM * 2;      (*MORE PAIRS PER CELL*)
     END;     (*CASE INVERSE*)
  END;     (*CASE*)
```

```
UNTIL (((INV=FWD) AND (CELDIS<2))
     OR ((INV=INVERSE) AND (CELNUM=0)));
END;(*SHUFFL*)
(***************************************************************************)
PROCEDURE FFT(INV:XFORM);
(*FAST FOURIER TRANSFORM PROCEDURE*)
(*OPERATES ON DATA IN 1ST HALF REAL 2ND-HALF IMAGINARY ORDER
AND PRODUCES A COMPLEX RESULT*)

VAR
     N1,N2,NU,CELNUM,CELDIS,PARNUM,IPN2,KPN2,JPN2,
     I,J,K,L,I2,IMAX,INDEX:INTEGER;
     ARG,COSY,SINY,R2COSY,R2SINY,I2COSY,I2SINY,PICONS,
     Y,DELTAY,K1,K2,TR,TI,XTEMP:REAL;


 BEGIN

(*CALCULATE NU = LOG2(N)*)

NU:=0;
N1:=N DIV 2;
N2:=N1;
WHILE N1>=2 DO
     BEGIN
     NU:=NU+1;          (*INCREMENT POWER-OF-TWO COUNTER*)
     N1:=N1 DIV 2;      (*DIVIDE BY 2 UNTIL 0*)
     END;
(*SHUFFLE THE DATA INTO BIT-INVERTED ORDER*)
FOR I:= 1 TO N2 DO
     BEGIN
     K:=IBITR(I-1,NU)+1; (*CALC BIT-INVERTED POSITION IN ARRAY*)
     IF I>K THEN          (*PREVENT SWAPPING TWICE*)
          (*INTERCHANGE I PT WITH ITS BIT-INVERTED MATE K*)
          BEGIN
          IPN2:=I+N2;
          KPN2:=K+N2;
          TR:=X[K];          (*TEMP STORAGE OF REAL*)
          TI:=X[KPN2];       (*TEMP IMAG*)
          X[K]:=X[I];
          X[KPN2]:=X[IPN2];
          X[I]:=TR;
          X[IPN2]:=TI;
          END;(*IF*)
     END;(*FOR*)

(*DO FIRST PASS SPECIALLY, SINCE IT HAS NO MULTIPLICATIONS*)
I:=1;
WHILE I<=N2 DO
     BEGIN
     K:=I+1;
     KPN2:=K+N2;
     IPN2:=I+N2;
     K1:=X[I]+X[K];         (*SAVE THIS SUM*)
     X[K]:=X[I]-X[K];       (*DIFF TO K'S*)
     X[I]:=K1;              (*SUM TO I'S*)
     K1:=X[IPN2] + X[KPN2];   (*SUM OF IMAG*)
     X[KPN2]:=X[IPN2]-X[KPN2];
     X[IPN2]:=K1;
     I:=I+2;
     END;     (*WHILE*)
```

```
            (*SET UP DELTAY FOR 2ND PASS  DELTAY= PI/2*)
                 DELTAY:=PI2;     (*PI OVER 2*)
                 CELNUM:=N2 DIV 4;
                 PARNUM:=2;       (*NUMBER OF PAIRS PER CELL*)
                 CELDIS:=2;       (*DISTANCE BETWEEN CELLS*)
(*EACH PASS AFTER 1ST STARTS HERE*)
REPEAT         (*UNTIL NUMBER OF  CELLS BECOMES ZERO*)


     (*EACH NEW CELL STARTS HERE*)

     INDEX:= 1;
     Y:=0;    (*EXPONENT OF W*)
     (*DO THE NUMBER OF PAIRS IN EACH CELL*)
     FOR I2:=1 TO PARNUM DO
          BEGIN
          IF Y<>0 THEN
               BEGIN              (*USE SINES AND COS IF Y IS NOT 0*)
               COSY:=COS(Y);      (*CALC SINE AND COSINE*)
               SINY:=SIN(Y);
               (*NEGATE SINE TERMS IF TRANSFORM IS INVERSE*)
               IF INV=INVERSE THEN SINY:=-SINY;
               END;(*IF*)
(*THESE ARE THE FUNDAMENTAL EQUATIONS OF THE FOURIER TRANSFORM*)
          FOR L:= 0 TO CELNUM-1 DO
               BEGIN
               I:=(CELDIS*2)*L + INDEX;
               J:=I+CELDIS;
               IPN2:=I+N2;
               JPN2:=J+N2;
               IF Y=0 THEN (*SPECIAL CASE FOR Y=0-- NO SINE OR COSINE TERMS*)
                    BEGIN
                    K1:=X[I]+X[J];
                    K2:=X[IPN2]+X[JPN2];
                    X[J]:=X[I]-X[J];
                    X[JPN2]:=X[IPN2]-X[JPN2];
                    END (*IF-THEN*)
               ELSE     (*IF Y<>0 USE SINE AND COSINE TERMS*)
                    BEGIN
                    R2COSY:=X[J]*COSY;   (*CALC INTERMEDIATE CONSTANTS*)
                    R2SINY:=X[J]*SINY;
                    I2COSY:=X[JPN2]*COSY;
                    I2SINY:=X[JPN2]*SINY;

               (*THESE ARE THE BASIC EQUATIONS OF THE FOURIER TRANSFORM*)
                    K1:=X[I] + R2COSY + I2SINY;
                    K2:=X[IPN2] - R2SINY +I2COSY;
                    X[J]:=X[I] - R2COSY-I2SINY;
                    X[JPN2]:=X[IPN2] + R2SINY -I2COSY;
                    END;(*ELSE*)
               (*REPLACE THE I TERMS*)
               X[I]:=K1;
               X[IPN2]:=K2;

     (*ADVANCE ANGLE FOR NEXT PAIR*)
          END;(*FOR L*)
     Y:=Y+DELTAY;
     INDEX:=INDEX+1;
     END;(*FOR I2*)
          (*PASS DONE- CHANGE CELL DISTANCE AND NUMBER OF CELLS*)
               CELNUM:=CELNUM DIV 2;
               PARNUM:=PARNUM * 2;
               CELDIS:=CELDIS * 2;
               DELTAY:=DELTAY/2;
          UNTIL CELNUM=0;
          END;(*FFT*)
```

```
(**************************************************************************)
BEGIN    (*MAIN PROGRAM*)
RESET(F1,'FTDAT DAT');   (*OPEN FILE FOR INPUT*)

(*READ IN FILE, A NUMBER AT A TIME*)
I:=0;
WHILE NOT EOF(F1) DO
     BEGIN
     I:=I+1;
     X[I]:=F1^;   (*GET EACH REAL NUMBER FROM FILE WINDOW*)
     GET(F1);
     END; (*WHILE*)
N:=I;         (*SAVE TOTAL READ IN*)

(*PERFORM THE SHUFFLE, FT AND POST-PROCESSING*)
SHUFFL(FWD);
FFT(FWD);
POST(FWD);
DEBUG;   (*WRITE OUT THE ANSWER FOR CHECKING*)
END.
```

EXAMPLE OF FOURIER TRANSFORM DATA

As an aid in entry and testing the Fourier transform routines, the following table indicates the calculated values for the forward real transform of a 16-point real array having the values 8, 16, 24, . . ., 128:

Table of Results of a Forward Real Transform

Initial	Shuffle	Bit invert	After 1st pass	After 2nd pass	After 3rd pass	After post-processing
8.00	8.00	8.00	80.00	224.00	512.00	512.00
16.00	24.00	71.99	−63.99	−128.00	−218.50	−127.99
24.00	40.00	40.00	143.99	−63.99	−128.00	−128.00
32.00	56.00	103.99	−63.99	−0.00	−90.50	−128.00
40.00	71.99	24.00	112.00	287.99	−63.99	−127.99
48.00	87.99	87.99	−63.99	−128.00	−37.49	−127.99
56.00	103.99	56.00	175.99	−63.99	0.00	−127.99
64.00	120.00	120.00	−63.99	0.00	90.50	−128.00
72.00	15.99	15.99	96.00	256.00	575.99	575.99
80.00	31.99	80.00	−63.99	−0.00	90.50	643.49

PROBLEMS

1 Write a program to generate a 4096-point, 100-cycle cosine wave and perform a forward real Fourier transform on it. Print out the index and contents of the five maximum array elements after the transform.

REFERENCES

1 P. R. Griffiths, "Chemical Infrared Fourier Transform Spectroscopy," Wiley-Interscience, New York, 1975.

2 T. C. Farrar and E. D. Becker, "Pulse and Fourier Transform Nmr," Academic Press, New York, 1971.

3 J. W. Cooper, "Spectroscopic Techniques for Organic Chemists," Wiley-Interscience, New York, 1980.

4 J. W. Cooley and J. W. Tukey, *Math. Comput.,* 19, 297 (1965).

5 E. O. Brigham, "The Fast Fourier Transform," Prentice-Hall, Englewood Cliffs, N.J., 1974.

6 G-AE Subcommittee on Measurement Concepts, W. T. Cochran et al., *IEEE Trans. Audio-electroacoust.,* AU15, 35–55 (1967).

7 J. W. Cooper, "Data Handling in Fourier Transform Spectroscopy," Chapter 4 of "Transform Techniques in Analytical Chemistry," P. Griffiths, editor, Plenum, New York, 1978.

8 J. W. Cooper, "The Minicomputer in the Laboratory: With Examples Using the PDP-11," Wiley-Interscience, New York, 1977.

9 J. W. Cooper, "Speeding up the Fast Fourier Transform for Nmr Data Processing," presented at the 21st Experimental Nmr Conference, Tallahassee, Fl., March 1980.

APPENDIX A

Reserved Words in Pascal

The following are Pascal reserved words or verbs that which cannot be used as type, constant, variable, procedure, or function names:

AND
ARRAY
BEGIN
CASE
CONST
DIV
DO
DOWNTO
ELSE
END
EXIT
EXTERN(AL)
FILE
FOR
FORWARD
FUNCTION
GET
GOTO
IF
IN
LABEL
MOD
NIL
NOT
OF
OR
PACKED
PROCEDURE
PROGRAM
PUT
RECORD
REPEAT
SET
THEN
TO
TYPE
UNTIL

VAR
WHILE
WITH

The following are the standard predefined Pascal functions:

Functions Returning a Real Value

ABS (X)	Absolute value of X
ARCTAN (X)	Arctangent of angle X (radians)
COS (X)	Cosine of angle X (radians)
EXP (X)	Exp (X)
LN (X)	Natural log of X
LOG (X)	Common log of X (not in all versions)
SIN (X)	Sine of angle X (radians)
SQR (X)	X squared
SQRT (X)	Square root of X

Functions Returning an Integer Value

ORD (C)	Integer value of character C
ROUND (X)	Rounds the real number and converts to integer
TRUNC (X)	Truncates the real number and converts to integer
ABS (I)	Absolute value of integer I
SQR (I)	Square
PRED (I)	Value of I − 1
SUCC (I)	Value of I + 1

Functions Returning a Character Value

CHR (X)	Returns the character equal to the integer value X
PRED (X)	Returns the previous alphabetic character
SUCC (X)	Returns the next alphabetic character

Functions Returning a Boolean Value

ODD (I)	Returns TRUE if I is odd
EOF (file)	Returns TRUE if scanning has reached an end of file
EOLN (file)	Returns TRUE if next character is a return from file or terminal

APPENDIX B

Pascal Punctuation Marks

The following punctuation marks have special uses in Pascal:

Semicolon	;	Used as a statement separator. Never used in the middle of IF-THEN-ELSE
Colon	:	Used between list of variable names and their type Used following Labels Used after case list constants Used preceding record variant lists Used to separate field length and fractional digit count in WRITE statements Used with "=" in assignment statements
Parentheses	()	Used with * in comments Used in definition of enumeration types Used to bracket arguments in procedure and function calls Used to define sets Used to enclose each variant part of Records Used to control order of evaluation in arithmetic expressions
Brackets	[]	Used in array indices Used to define array ranges Used in constant set constructions
Equals	=	Used to define types Used with : in assignment statements Used to define constants Used in relational Boolean expressions
Asterisk	*	Used for multiplications Used with (and) in comments Used in set intersection
Slash	/	Used for division of real numbers
Up-arrow	↑	Used to denote a pointer variable Precedes variable in declaration Follows variable name in statement
Period	.	Follows END of program Record component selection Used as decimal point

Comma	,	Separated elements in lists, set constructions, array indices, parameter declarations, and parameter lists
Apostrophe	'	String delimiter—two used inside string
Elipsis	..	Used to separate lower from upper bound of a subrange type or array dimension
Plus	+	Addition Set union
Minus	−	Subtraction Negation Set difference
Relational Operators	>	Greater than
	<	Less than
	< >	Not equal
	< =	Less than or equal to
	> =	Greater than or equal to

APPENDIX C

Using Pascal on the DEC-10

To use the Pascal compiler on the DEC-10, your program files should all have the ".PAS" extension. To compile and execute your program, you need only give the EXecute command followed by the file name of your program. The extension ".PAS" need not be typed. To execute the file SUMMER.PAS, type:

```
.EX SUMMER
```

The compiler will then print out:

```
PASCAL: SUMMER    [SUMMER   ]

   0 ERRORS DETECTED        (if none were)

HIGHSEG:    OK + nnn WORD(S)
LOWSEG :    OK + nnn WORD(S)

RUNTIME:    0: N.NNN

LINK:    Loading              (here the compiled file is loaded
[LNKXCT SUMMER  execution]      and executed)
```

and then begin executing the program.

Upon exit from the program, you will find a file in your area named SUMMER.REL, the compiled file, and SUMMER.LST, the listing file. The listing file contains the listing of the compiled program with line numbers and core occupied. It is not generally too useful, except in long programs or in ones where error messages occur.

Error messages are printed out on the terminal for each error found during compilation and are also made part of the listing file under the incorrect line. The most obscure and confusing messages usually result if the *previous* line does not contain any semicolon.

SPECIAL FEATURES AND RESTRICTIONS OF DEC-10 PASCAL

In DEC-10 Pascal, CMU version, there are some special features and restrictions:

1 Single characters are not read from the terminal until the line containing those characters is terminated with a return. If the character is read with a READ, then the return is still the next character, but if the character is read with a READLN, the return is skipped over as usual.

2 The CASE statement does not skip over all cases if no match is found (in accordance with the proposed standard), and the OTHERS selector can be used for all values not specified as specific selectors.
3 The index variable of the FOR statement must be a variable declared in that procedure, or the compiler will issue a warning message. (This is also in accordance with the proposed standard.)
4 The run-time system does not automatically expand inadequate formats but prints a pair of asterisks instead.
5 The file variable INPUT refers to the terminal and may alternatively be referred to as TTY for compatibility with older versions.
6 Invalid format REAL or INTEGER data entered from the terminal cause an error message, and entry of new values is then allowed.
7 Attempts to read beyond the end of a text file will be allowed only eight times, and then a run-time error will occur. Attempts to read a real or integer number beyond the end of a file will always cause a run-time error, since these read routines skip over leading blanks. Thus they will still be looking for numbers after eight tries, and a run-time error will occur.

LINKING EXTERNALLY COMPILED PROGRAMS WITH PASCAL

Linking of externally compiled programs is extremely easy in Pascal. The procedures to be linked must be declared as EXTERN in the main program, with all their parameters declared as usual:

```
PROCEDURE OUTSIDE (VAR X:REAL; K:INTEGER);
EXTERN;
```

and then the main program is compiled as usual. The separately compiled procedure .REL file must also be present, and the EX command should then include the main program filename, the procedure filename, and the Pascal library:

```
.EX MAINPR,OUTSID,PASLIB/LIB
```

PECULIARITIES OF TERMINAL INPUT IN DEC-10 PASCAL

The following discussion applies to the Kisicki and Nagel modifications of the Hamburg compiler on the DEC-10. It does not apply to the CMU compiler used throughout this book.

In DEC-10 Pascal the terminal is a separate input channel from the standard file INPUT but is treated just as other input character files are, including the look-ahead file window feature. However, there are some specific peculiarities and drawbacks to this version of Pascal.

Specifically, all input and output to the terminal must be done using the file identifier TTY as the first argument to any READ or WRITE statement:

```
READ(TTY,A,B); (*READ A AND B FROM THE TTY*)
WRITELN(TTY, A:10, B:12);   (*WRITE A AND B
                             ON THE TERMINAL*)
```

It is not possible to read any character, string, or number from the terminal without terminating a group of such numbers with a return, since it is only then that the data is passed from the DEC-10 terminal line buffer to the Pascal program. Further, while the return character will be the *next* character after the data is read, a READLN(TTY,list) will not be satisfied with that terminating return since the look-ahead character after the return will not yet be read. Thus the READLN statement can never be used in conversational programming from the TTY, since Pascal will wait for still another input character beyond the return. In addition, this additional input character is not passed to Pascal until another return is typed, since only lines are passed to Pascal, not characters. Thus the statement

```
READLN(TTY,A,B);
```

will wait for a *second* return after reading A and B from the line passed by the first return. To circumvent this clumsiness in conversational programming with DEC-10 Pascal, simply use the READ statement in all cases, rather than the READLN statement when reading from the TTY.

DEC-10 Character Input

The DEC-10 Pascal system automatically puts a character in the look-ahead window of the file TTY when any program starts that will utilize READs from the terminal. At the time any DEC-10 Pascal program is EXecuted from the monitor, the program is linked and loaded and types an asterisk to indicate that it is ready for that first character of input. This is usually just a return (which, of course, becomes a space when passed to Pascal).

This space in the file window is usually irrelevant if the first data to be entered from the terminal is to be a number of type INTEGER or REAL, since leading spaces are always ignored. However, if the first data in the program is to be a character input from the terminal, it will be that space in the look-ahead buffer unless this space is read and discarded. Consider the following:

```
PROGRAM CHTEST;
VAR A,B,C:CHAR;
(*THIS PROGRAM ILLUSTRATES THE INCORRECT
USE OF THE LOOK-AHEAD CHARACTER *)

BEGIN
READ(TTY,A,B,C);      (*A WILL BE SET TO LOOK-AHEAD CHAR*)
WRITELN(TTY, 'A=',A);
WRITELN(TTY,'B=',B);
WRITELN(TTY,'C=',C);
END.
```

Let us assume that the user runs this program by typing

EX CHTEST.PAS

and, when the program is loaded, it types

*

The user then types a return followed by

ABC

The output of the program will then be

A=
B=A
C=B

because the TTY window was loaded with the first character, a space, which was then read into variable A. Then variables B and C were loaded with the characters A and B, respectively. In any program where the *first* data to be read in will be a character, you must read and "throw away" that extra window character which is entered first:

```
PROGRAM CHTEST2;
VAR A,B,C,DUMMY:CHAR;
(*THIS PROGRAM ILLUSTRATES THE
CORRECT USE OF THE LOOK-AHEAD CHARACTER*)

BEGIN
READ(TTY,DUMMY);      (*THROW AWAY FIRST CHARACTER*)
READ(TTY,A,B,C);      (*NOW READ A,B,C CORRECTLY*)
WRITELN(TTY, 'A=',A);
WRITELN(TTY,'B=',B);
WRITELN(TTY,'C=',C);
END.
```

In addition, if the next data to be read in is a character and the previous data entered and read was terminated with a return, this return will still be present as a space which you must read and throw away.

Another difficulty of conversational Pascal programming on the DEC-10 occurs when you wish to use a single character command to decide what calculation to perform next. The program fragment

```
WRITE(TTY,'ENTER ONE CHARACTER: '); BREAK(TTY);

READ(TTY,C);          (*READ ONE COMMAND CHARACTER*)
```

will not continue executing until both the single character *and* the return are typed. Since it is so common in conversational programming to terminate such commands with a return, this is seldom a problem, however, but should be recognized clearly by the programmer as being nonstandard Pascal.

APPENDIX D

Using Pascal on the Aspect-2000

To run Pascal on the Aspect-2000, the files COMPILER.CODE, PASERR, PASCAL, and LINKER.CODE must be present on the disk and the ADAKOS monitor installed and running.

Source files are created using the TECO editor and can be compiled using the command:

```
RUN PASCAL filename:C
```

To compile and produce a listing, the command

```
RUN PASCAL filename:CL
```

should be given, and to send the listing to the high-speed printer, the command

```
RUN PASCAL filename:CLP
```

should be given.

All files created by TECO have an .ASC extension which need not be specified in the RUN command, and the compiled p-code files have a .CODE extension.

To execute the compiled program, the command

```
RUN PASCAL filename
```

should be given, without any extension or modifiers after a colon. If program output is to go to the high-speed printer, the command

```
RUN PASCAL filename:P
```

should be given.

APPENDIX E

Using UCSD Pascal

Structure of Pascal Diskettes
Starting up UCSD Pascal
Segmenting Procedures
CASE Statements
Interactive Files
The GOTO Statement
Untyped Files

This section describes the use of UCSD Pascal on the PDP-11. Minor differences occur with other computers' implementations of UCSD Pascal, but the command structure and compilation process are the same.

STRUCTURE OF PASCAL DISKETTES

UCSD Pascal runs under its own disk operating system which is distinct from RT-11, although it is possible to have Pascal diskettes which can be read by RT-11 V02C. The Pascal directory occupies tracks 2–5, with a duplicate directory on tracks 6–9 being optional at the time each disk is zeroed. RT-11 directories occupy tracks 6–9, and thus a disk which is to be read by both Pascal and RT-11 cannot have a duplicate Pascal directory. RT-11 V3 bootstraps cannot be written onto Pascal disks, but V3 is otherwise compatible with Pascal.

STARTING UP UCSD PASCAL

To start up UCSD Pascal, place the disk PASCALT in the left-hand diskette drive and any disk in the right-hand drive, and bootstrap the system in the usual way from the hardware or software bootstrap. It is important that there be a disk in both drives, or the Pascal bootstrap may hang up.

The program will print out

```
WELCOME, PASCALT, TO UCSD PASCAL
```

and print a command prompt line:

```
Command: E(dit, R(un, F(ile, C(omp, L(ink, X(ecute, A(ssem, D(ebug,
?
```

or, if the system has a slow terminal, and has been so configured:

```
Command:
```

The E command enters whatever editor is named as SYSTEM.EDITOR, which may be either a screen or a line-oriented editor, and allows creation of source files on either disk. The F command allows zeroing of blank or old disks and copying of files. The C command specifically calls for the compilation of the file SYSTEM.WRK.TEXT into SYSTEM.WRK.CODE, and if no such file currently exists on the system disk, the program prompts you for a file name to compile. The X command executes the file SYSTEM.WRK.CODE, and if no such file exists, compiles SYSTEM.WRK.TEXT and executes it. It, too, will prompt for a filename if no WRK files exist.

UCSD Pascal differs from standard Pascal primarily in the fact that the type STRING is defined as being a packed array of up to 80 characters with the string length in the zeroth byte. Thus strings can have any length and can be compared, even if of different lengths.

SEGMENTING PROCEDURES

Since UCSD Pascal is designed for use on minicomputers and 48K (byte) microcomputers, it allows procedures to be *segmented* and called into memory as needed. This is done by using the statement:

```
SEGMENT PROCEDURE GLARCH (xxx);
```

and then writing the procedure in the usual way. All of the procedures following the SEGMENT PROCEDURE are in that segment until the next SEGMENT is declared. UCSD Pascal allows up to seven segments for user programs.

CASE STATEMENTS

In UCSD Pascal, CASE statements do not issue a run-time error if the case variable is not one that is included in one of the cases. Instead, it simply drops through to the first statement after the CASE block.

INTERACTIVE FILES

To allow for interactive programming at the keyboard, UCSD Pascal defines the file type INTERACTIVE, which is used for the default input and output file, the keyboard. Here the look-ahead character INPUT↑ does not really exist. The procedure READ(CH) for a character expands to

```
GET(F);
CH:=F^;
```

where F is a file of type INTERACTIVE and CH is a character. This differs of course for the standard definition for READ(F1,C), which is

```
CH:=F^;
GET(F); (*LOOK-AHEAD CHARACTER IN WINDOW*)
```

THE GOTO STATEMENT

The GOTO statement causes a compile-time error message unless the compiler comment (*$G+*) is given.

UNTYPED FILES

UCSD Pascal also allows the reading and writing of untyped files having no associated file window, using the BLOCKREAD and BLOCKWRITE statements. Consult the UCSD Pascal manual for further details.

APPENDIX F

Limitations of Pascal

No Run-Time Dimensioning
Lack of Double Precision
Random Number Generations
Lack of Random Access to File
No Standard for Interactive Systems
Lack of Exponentiation
Lack of Procedures Passed as Parameters
Lack of Constant Expressions

While Pascal is in many ways an ideal language for both scientific and general use, it does have a few limitations which we must point out.

NO RUN-TIME DIMENSIONING

The size of all arrays must be declared in advance in Pascal, rather than at run time. Thus a program must be compiled with the largest size array you expect to use, and recompiled if this size changes in either direction.

Some versions of Pascal allow you to subvert this with certain stack manipulations using NEW, MARK, and RELEASE statements, but these are awkward, must be referred to through pointers, and are not standard Pascal.

LACK OF DOUBLE PRECISION

While many features can be programmed within Pascal itself, it is not possible to carry out double precision calculations since standard Pascal compilers cannot generate instructions to carry out these operations.

Double precision is unnecessary on machines with extremely long word lengths, such as the CDC Cyber series, but may be quite important in scientific calculations in machines having shorter word lengths, such as the DEC-10 in some cases, the IBM computers, and most minicomputers.

At present double precision can only be carried out by declaring all double precision numbers as a RECORD of two words length, and calling externally linked assembly language routines for all double precision operations.

RANDOM NUMBER GENERATORS

Standard Pascal does not include a random number generator for pseudo-random numbers, although many systems have added this feature as a callable function. Consult local documentation for details.

LACK OF RANDOM ACCESS TO FILES

Standard Pascal does not provide any way to read or write an arbitrary record from a file, as would be handy in updating various sorts of record-keeping programs. Many versions have added such features, but they differ from installation to installation.

NO STANDARD FOR INTERACTIVE SYSTEMS

Pascal does not set any standard for systems that run Pascal in a conversational mode. Thus whether a READ of a character requires that the return be typed to pass a line buffer to the program is implementation-dependent. Further the state of the terminal look-ahead character is quite different with different interactive systems, some even requiring that all extra returns and spaces be entered on following lines before the READ is complete.

LACK OF EXPONENTIATION

There is no exponentiation operator in Pascal, although there is a serious effort to have one included in a future standard, since the logarithmic form is cumbersome to use and easy to confuse. One suggestion is a predefined function power:

```
FUNCTION POWER (A,B:REAL):REAL;
```

which calculates the *b*th power of *a*. Other users prefer the common '**' operator used in FORTRAN.

LACK OF PROCEDURES PASSED AS PARAMETERS

Most Pascals today have not implemented the ability to pass a procedure or function by name to a procedure as an argument. This definitely limits the language.

LACK OF CONSTANT EXPRESSIONS

Pascal does not allow you to compute constants within the program text, but requires them to be entered in their final form. Thus

```
CONST
   PIOVER2=3.14159/2;
```

is not allowed.

APPENDIX G

The Pascal User's Group and Newsletter

The Pascal User's Group (PUG) promotes the use of the Pascal programming language as well as the ideas behind Pascal through the publication *Pascal News.* PUG is intentionally designed to be nonpolitical and thus does not "take stands" on issues or causes, however well intentioned. The organization is entirely informal; there are no officers or meetings of PUG.

Anyone can join PUG, particularly the Pascal user, teacher, maintainer, implementor, distributor, or fan. Library memberships are also encouraged. To obtain information and subscribe to *Pascal News,* contact one of the following locations. The 1980 prices are indicated in parentheses:

Pascal User's Group, c/o Rick Shaw ($6.00)
Digital Equipment Corporation
5775 Peachtree Dunwoody Rd.
Atlanta, Georgia 30342, USA

Pascal User's Group (£4.00)
c/o Computer Studies Group
Mathematics Department
The University
Southamption S09 5NH, United Kingdom

Pascal User's Group, c/o Arthur Sale ($A8.00)
Department of Information Science
University of Tasmania
Box 252C GPO
Hobart, Tasmania 7001
Australia

Pascal News publishes comments and information from all Pascal users, and all are encouraged to communicate their experiences and/or frustrations. Material should be single spaced and camera ready, typed with a dark ribbon with lines 18.5 cm wide.

Answers to Selected Problems

CHAPTER 1

1 A compiler translates complex statements into many lines of machine code.

2 A portable compiler translates complex statements into some sort of code that can be executed or emulated by a number of different kinds of computers.

3 A one-pass compiler requires much less compilation time since the program file must only be scanned once. It also may result in imposing greater structure and organization on the programmer, leading to less errors in the program.

CHAPTER 2

1 Pascal would evaluate the expression as
X:= 5 + ((4*6)/7) which equals 8.4285.

2 The LN of 5.0E7 is 17.7275, and Y will be the integer less than that value, or 17.

3 The following may be illegal:

BEGIN; An empty statement after BEGIN is at least meaningless.

5:=SQR(2.2); Can't assign a new value to a constant

(* COMMENT! *) Can't have a space between the star and the parenthesis in a comment.

WRITELN(*A,B*); Not illegal, but certainly confusing. The statement is equivalent to WRITELN; since the apparent arguments are actually enclosed in a comment.

CHAPTER 3

1 X:=EXP(11*LN(5));

2 Let us assume that this is all one program in which all the constants and variables must be predeclared and separators are required:

VAR X,Y:REAL is illegal because it does not end with a semicolon.

The statement M:=X*Y; would only be legal if X and Y were of type INTEGER. Otherwise the function BOUND or TRUNC would have to be used.

The up-arrow symbol cannot be used for raising to a power in Pascal. Instead, the expression should be written Y:=M*X + SQR(B); However, the expression is still illegal because B is of type BOOLEAN in this program and cannot be squared.

CK:=M+N; cannot be allowed because a character variable is being set to an integer value. This should be written CK:=CHR(M+N);

Likewise, X:=Y+CK; cannot be allowed unless the character variable is converted to a number: X:=Y+ ORD(CK);

3 A WRITE statement could be written on two lines if the string were broken into two parts:

WRITE ('FOURSCORE AND SEVEN YEARS AGO',
'OUR FATHERS . . .');

4 B:=M < >N;

5
```
(*PROBLEM 3.5*)
PROGRAM SPHERE;
(*CALCULATES THE VOLUME OF A SPHERE:
                3
V = 4/3 PI R   *)

CONST F3PI=4.188790205;
VAR RADIUS,VOLUME:REAL;

BEGIN
   WRITE('ENTER RADIUS: ');      (*ASK FOR RADIUS*)
   READLN(RADIUS);       (*GET IT FROM THE TERMINAL*)

(*CALCULATE THE VOLUME*)
   VOLUME:=F3PI*EXP(3*LN(RADIUS));
   WRITELN('VOLUME OF SPHERE =',VOLUME);
END.
```

6 G(V):=GMAX/(1.0 +SQR (2*(V−V0)/LINEWIDTH));

CHAPTER 4

1 DPOS:=''''; character type, a single apostrophe

PI=3.1416; real

FOUR=4; integer

NAME='NAUGHTYBITS'; string

A='A'; character

NINE=9.0; real

2 CONST, 5FOLD and "SMILE' are illegal. BEGINS, IN1T and TO0 are unwise.

CHAPTER 5

2
```
(*PROBLEM 5.2
READ IN 10 NUMBERS AND PRINT OUT
SMALLEST NON-ZERO VALUE *)

CONST   MAX=10;

VAR
    I:INTEGER;
    MIN:REAL;
    X:ARRAY[1..MAX] OF REAL;

BEGIN
(*READ IN 10 NUMBERS*)
FOR I:=1 TO MAX DO READLN(X[I]);

(*FIND SMALLEST NON-ZERO VALUE*)
MIN:=1.0E35;     (*SET MIN VERY LARGE*)
FOR I:=1 TO MAX DO
    IF (X[I]<MIN) AND (X[I]<>0.0 ) THEN
    MIN:=X[I];
WRITELN('SMALLEST NON-ZERO VALUE = ',MIN);
END.
```

3
```
(*PROBLEM 5.3*)

PROGRAM TEMPERATURE;
(*READS IN A NUMBER AND 'C' OR 'F'--
IF F, CONVERTS TO CELSIUS
IF C, CONVERTS TO FAHRENHEIT *)

VAR F,C,X:REAL;
COMD:CHAR;

BEGIN
WRITE('ENTER TEMP: ');
    READLN(X);   (*GET TEMPERATURE*)
WRITE('F OR C: ');
    READLN(COMD);

(*NOW EXCLUDE ALL CHARS  BUT F OR C *)
IF (COMD='C') OR (COMD='F') THEN
CASE COMD OF
    'F':     BEGIN
        F:=X;    (*SET FAHRENHEIT TEMP*)
        C:=(F-32)*5/9;
        WRITELN(F,' FAHRENHEIT= ',C,' CELSIUS');
        END;
```

```
      'C':     BEGIN
          C:=X;    (*SET CELSIUS*)
          F:=(9*C)/5 + 32;
          WRITELN(C, ' CELSIUS= ',F, ' FAHRENHEIT');
          END;
     END;      (*CASE*)
 END.
```

5
```
(*PROBLEM 5.5*)
PROGRAM ZBLOCK;
(*READS IN THE DATA IN A 4096-WORD FILE
AND PRINTS OUT THE SIZE AND LOCATION
OF THE LARGEST BLOCK OF SUCCESSIVE ZEROS FOUND*)

CONST MAX=12;

VAR
    X:ARRAY[1..MAX] OF REAL;
    I,ZMIN,ZMAX,NZMIN,NZMAX,ZSIZE:INTEGER;
    F1:TEXT;     (*FILE CONTAINING DATA TO READ*)

BEGIN
RESET(F1,'ZDATA     ');   (*DATA FILE*)
    FOR I:=1 TO MAX DO READ(F1,X[I]);
(*INITIALIZE CONSTANTS *)
ZMIN:=0;     ZMAX:=0,     NZMIN:=0;    NZMAX:=0;    ZSIZE:=0;
FOR I:=1 TO MAX DO
    BEGIN
    IF X[I]=0 THEN
        IF NZMIN=0 THEN NZMIN:=I  (*REMEMBER START OF ZERO BLOCK*);
(*TERMINATE COUNTING OF ZEROS IF
    NEXT ENTRY IS NOT ZERO
    OR IF  THIS IS LAST ELEMENT OF ARRAY *)
    IF ((X[I]<>0) AND (NZMIN <>0)) OR (I=MAX) THEN
        BEGIN
        IF I<>MAX THEN NZMAX:=I-1
            ELSE NZMAX:=I;
        IF(NZMAX-NZMIN)>ZSIZE THEN
            BEGIN
        (*COPY SIZE AND BOUNDS OF NEWEST BLOCK
        TO ZMIN, ZMAX AND ZSIZE IF THE NEWEST BLOCK
        IS GREATER THAN THE LAST ONE *)
            ZMIN:=NZMIN;
            ZMAX:=NZMAX;
            ZSIZE:=ZMAX-ZMIN+1;
            NZMIN:=0;   (*ZERO MEANS LOOK FOR NEW BLOCK*)
            NZMAX:=0;
            END; (*IF NZMAX*)
        END;(*IF X*)
    END;(*FOR*)
(*PRINT OUT LARGEST BLOCK*)
WRITELN('LARGEST BLOCK SIZE=',ZSIZE,' FROM ',ZMIN,' TO ',ZMAX);
END.
```

6 The value 128 is, of course, a power of 2 and can be represented exactly in binary, but the value 12.8 is not, and can only be approximated. Multiplying this approximate value by 10 will produce approximately 128, and perhaps not exactly 128.

CHAPTER 6

1
```
(*PROBLEM 6.1*)
FUNCTION LINE(M,X,B:REAL):REAL;
BEGIN
    LINE:=M*X+B;
END;(*LINE*)
```

3
```
(*PROBLEM 6.3*)
PROCEDURE MAXDIV(VAR X:ARRAY[1..TOP,1..TOP] OF REAL);
BEGIN
MAX:=-1.0E35;    (*LARGE NEGATIVE  NUMBER*)
(*LOOK FOR LARGEST VALUE IN ARRAY*)
FOR I:=1 TO TOP DO
    FOR J:=1 TO TOP DO
    IF MAX<X[I,J] THEN MAX:=X[I];
(*DIVIDE THROUGH BY LARGEST VALUE FOUND*)
FOR I:=1 TO TOP DO
    FOR J:=1 TO TOP DO
    X[I,J]:=X[I,J]/MAX;
END;     (*MAXDIV*)
```

5
```
(*PROBLEM 6.5*)
PROGRAM LCFDO;
(*THIS PROGRAM CALCULATES THE LARGEST COMMON FACTOR
OF TWO ENTERED NUMBERS*)

VAR X,Y,Z:INTEGER;

FUNCTION LCF(A,B:INTEGER):INTEGER;
(*PROCEDURE CALCULATES THE LARGEST COMMON FACTOR
RECURSIVELY. RETURNS WITH A IF B=0,
OTHERWISE CALLS LCF AGAIN WITH A MOD B *)
BEGIN
IF B=0 THEN LCF:=A
ELSE
LCF:=LCF(B,A MOD B);
END;(*LCF*)

BEGIN
WRITE('ENTER X Y ');
READLN(X,Y);
Z:=LCF(X,Y);     (*CALCULATE LARGEST COMMON FACTOR*)
WRITELN(Z,' IS THE LARGEST COMMON FACTOR OF ',X,' AND',Y);
END.
*
```

CHAPTER 7

1
```
(*PROBLEM 7.1 *)
PROGRAM CHARREV;
(*READS IN A STRING AND PRINTS IT OUT IN REVERSE ORDER*)

CONST LINMAX=80;     (*MAX IN ONE LINE*)

VAR
I,LASTC:INTEGER;
C:ARRAY[1..LINMAX] OF CHAR;

BEGIN
I:=1;    (*INDEX OF CHARACTER ARRAY*)
REPEAT
    READ(C[I]);
```

```
    I:=I+1;
UNTIL (I>LINMAX) OR EOLN(INPUT);
LASTC:=I-1; (*REMEMBER # OF LAST CHARACTER*)
WRITELN;    (*SKIP A LINE*)
FOR I:=LASTC DOWNTO 1 DO     (*IN REVERSE ORDER*)
    WRITE(C[I]);
WRITELN;
END.
```

4

```
(*PROBLEM 7.4*)
PROGRAM DIAGNUM;
(*PRINTS OUT NUMBERS FROM 1.1 TO 9.9 IN A DIAGONAL LINE*)

CONST F=4;  D=1;    SP=' ';

VAR
I,SPACER:INTEGER;
X:REAL;

BEGIN
X:=1.1;
SPACER:=0;  (*START AT LEFT*)
REPEAT
    I:=0;
    WHILE I<SPACER DO   (*PRINT LEADING SPACES IN EACH ROW*)
        BEGIN
        WRITE(SP);  (*A SPACE AT A TIME*)
        I:=I+1; (*COUNT THEM*)
        END;(*WHILE*)
    WRITELN(X+0.05:F:D);    (*WRITE OUT CURRENT NUMBER*)
    X:=X+1.1;   (*ADD ON TO X*)
    SPACER:=SPACER+5;   (*MOVE OVER FOR NEXT #*)
UNTIL X> 10;    (*ONLY ALLOW NUMBERS UP TO 9.9*)
END.
```

CHAPTER 9

```
(*PROBLEM 9.1*)
PROGRAM ACLCNT;
(*COUNTS THE FREQUENCIES OF THE CHARACTERS 'A', 'C' AND 'L'
IN THE FILE 'PASCAL.RNM'
WHICH IS THE SOURCE OF THIS ENTIRE MANUSCRIPT*)

VAR
LCNT,CCNT,ACNT:INTEGER;
F1:TEXT;    B:CHAR;

BEGIN
RESET(F1,'PASCALRNM');  (*OPEN FILE*)
REPEAT
    READ(F1,B); (*READ A CHARACTER AT A TIME*)
    IF ((B='A') OR (B='C')) OR (B='L') THEN
    CASE B OF
```

```
        'A' :ACNT:=ACNT+1;  (*COUNT A'S*)
        'C':    CCNT:=CCNT+1;   (*C'S*)
        'L':    LCNT:=LCNT+1;   (*OR L'S*)
    END;(*CASE*)
UNTIL  EOF(F1);

(*THEN WRITE OUT THE ANSWERS*)
WRITELN(ACNT, ' A''s');
WRITELN(CCNT, ' C''s');
WRITELN(LCNT,' L''s');
END.
```

This manuscript contained 4209 A's, 2163 C's, and 3223 L's, before editing. Note that this only counts the capital letters.

CHAPTER 10

1

```
(*PROBLEM 10.1*)
PROGRAM SETS5;
(*THIS PROGRAM ALLOWS ENTRY OF 5 NUMBERS BETWEEN
1 AND 20 AND THEN TESTS FURTHER NUMBERS TO SEE
IF THEY ARE A MEMBER OF THAT SET ALREADY ENTERED*)

(*DEFINE SET TYPE*)
TYPE
    SUBINT=1..20;    (*SUBRANGE TYPE*)
    SETINT=SET OF SUBINT;    (*SET TYPE*)

VAR
    ASET:SETINT;     (*SET VARIABLE*)
    A:SUBINT;    (*INTEGER IN RANGE OF 1..20*)
    I:INTEGER;

(*****************************************************)
PROCEDURE GETSUBINT(VAR B:SUBINT);
(*PROCEDURE TO GET A NUMBER IN THE SUBRANGE 1..20
WITHOUT  PASCAL SYSTEM ISSUING ERROR MESSAGE
IF IT IS NOT. NUMBER IS FIRST ENTERED AS AN INTEGER
AND THEN CHECKED FOR BOUNDS BY PROCEDURE.*)

VAR X: INTEGER; (*LOCAL TEMPORARY VARIABLE*)
    INSET:BOOLEAN;  (*FLAG FOR SUCCESSFUL VALUE*)

BEGIN
INSET:=FALSE;    (*FLAG TO SEE IF IN RANGE*)
REPEAT
    WRITE('> ');     (*PROMPT CHARACTER*)
    READLN(X);  (*READ IN INTEGER*)
    (*CALCULATE WHETHER IN SUBRANGE*)
    INSET:=(X>=1) AND (X<=20);
    IF NOT INSET THEN WRITELN('ILLEGAL VALUE (1<X<20), RE-ENTER');
UNTIL INSET;     (*KEEP IT UP TILL WE GET A GOOD ONE*)
B:=X;    (*RETURN GOOD VALUE IN B*)
END;     (*GETSUBINT*)
(*****************************************************)
```

```
BEGIN   (*MAIN*)
ASET:=[];    (*INITIALIZE EMPTY SET*)
WRITELN('ENTER 5 NUMBERS BETWEEN 1 AND 20');
FOR I:=1 TO 5 DO
     BEGIN
     GETSUBINT(A);    (*READ THEM*)
     ASET:=ASET+[A]; (*COMBINE BY SET UNION*)
     END;
(*GET AND COMPARE NUMBERS UNTIL 5 WRONG ENTRIES MADE*)
WRITELN('NOW ENTER NUMBERS TO SEE IF THEY ARE IN THE SET');
I:=0;    (*COUNTS WRONG ENTRIES*)
REPEAT
     GETSUBINT(A);    (*READ EACH NUMBER*)
     IF A IN ASET THEN WRITELN(A,' IS A MEMBER')
     ELSE
     BEGIN
     WRITELN(A, ' IS NOT IN THE SET');
     I:=I+1;
     END;
UNTIL I>=5;

END.
```

CHAPTER 15

1 This is the classic example of confusing *value* parameters with *reference* parameters. The procedure GETJ has the argument (I:INTEGER) which it can refer to by *value* but cannot change. This should be rewritten as a reference parameter:

PROCEDURE GETJ(VAR I:INTEGER);

to make the program work properly.

Index